Classical Mechanics
Illustrated by Modern Physics

42 Problems with Solutions

Classical Mechanics
Illustrated by Modern Physics

42 Problems with Solutions

David Guéry-Odelin • Thierry Lahaye

Paul Sabatier University, France

ICP

Imperial College Press

Published by

Imperial College Press
57 Shelton Street
Covent Garden
London WC2H 9HE

Distributed by

World Scientific Publishing Co. Pte. Ltd.
5 Toh Tuck Link, Singapore 596224
USA office: 27 Warren Street, Suite 401-402, Hackensack, NJ 07601
UK office: 57 Shelton Street, Covent Garden, London WC2H 9HE

British Library Cataloguing-in-Publication Data
A catalogue record for this book is available from the British Library.

ISBN-13 978-1-84816-479-6
ISBN-10 1-84816-479-3
ISBN-13 978-1-84816-480-2 (pbk)
ISBN-10 1-84816-480-7 (pbk)

Printed in Singapore by World Scientific Printers.

Foreword

Among all subfields of physics, the most familiar to each of us is undoubtedly classical mechanics. Archimedes floating in his bath, Newton watching the fall of an apple, Galileo muttering "and yet it moves" have become universal characters. Their discoveries are part of common wisdom, and accepted by everyone as natural and fundamental rules.

But can we make sure that we really understand mechanics? Basic principles are written in a few lines and their simplicity is amazing. However, their domain of applicability is huge, ranging from molecular motion to galaxy dynamics. In order to master mechanics in all its diversity, one thus needs to use it; in academic terms, one needs to solve problems. But which ones? Facing traditional exercises, the reaction of students is most often: "the calculations are too heavy!" However, the examiner is often kind enough to add at the very end of the problem the recommendation "Interpret the result physically". But this is not sufficient to unveil to the students the appeal of hypothetical cones rolling inside vibrating cylinders.

The reflection themes proposed by David Guéry-Odelin and Thierry Lahaye are the complete opposite of the usual stereotyped exercises. In this book the readers will encounter particles, atoms, or planets, evolving in simple force fields such as gravity or Coulomb repulsion. In each text, the authors start with a well-defined physical question and guide their readers by estimating relevant orders of magnitude. Of course, calculations are not absent — studying the stability of Lagrange points implies some suffering — but they are never pointless. Above all, the authors show a true delight in bringing us to the discovery of hidden aspects of the dynamics of simple objects: a ball can stay in stable equilibrium on a saddle, atoms can be cooled when illuminated by laser beams, non-dissipative motion can occur in the presence of friction forces, and so on.

Classical mechanics is a lively subject. Many problems encountered in the present book refer to discoveries that occurred during the last two decades. The readers will thus learn mechanics at its best level, which is important for short-term academic success, and will discover pieces of contemporary research, which may be more important in the longer term for career preparation. The authors have managed to reach this double goal because they are both teachers and actors in this scientific adventure. Several chapters refer to their research field, and some are directly inspired by their own work. I sincerely hope that the passion that David Guéry-Odelin and Thierry Lahaye have put into this book will stimulate the curiosity of their readers, and teach them the pleasure of understanding how, beyond formalism, new physical phenomena can emerge.

Jean Dalibard
Researcher at CNRS
Professor at École Polytechnique

Preface

In many research fields, both in fundamental and applied physics, classical mechanics plays, even today, a key role. However, the traditional way of teaching mechanics often confines applications to old-fashioned pendulums, springs, or rotating cylinders. The goal of this book is to illustrate classical mechanics mostly by examples arising from *contemporary physics* — although we have included a few standard problems that cannot be ignored. The exercises and problems that are gathered here are the result of our teaching experience, in lectures, exercise classes, and written and oral exams.

This book is written mainly for undergraduate physics students. They will find material to test their understanding of physical concepts taught in lectures, to improve their ability to solve concrete problems, and also a way to discover some aspects of modern physics which should prove useful in the rest of their curriculum (for instance, links with statistical mechanics or quantum physics are emphasized when relevant). We have chosen to give complete solutions, with explicit calculations and many physical comments, so that the book can be used by the students on their own. In the same spirit, we recall, for most problems, basic notions in the form of "reminders". We hope that instructors will find here some useful material that can be adapted for the preparation of homework problem sets or exams.

We have given an indication of the level of difficulty of the exercises. In principle, the easiest ones (*) can be solved by first-year university students. Those of intermediate difficulty (**) require a deeper knowledge of classical mechanics. A few exercises and problems (***) have a more technical character. The reader may work on these problems using the solutions in order to widen his/her knowledge of classical mechanics but also of modern physics.

The themes we have chosen illustrate our personal interests, and, for a number of them, have been inspired directly by our research field (in particular the problems of Chapter 9), but also by discussions with colleagues or by articles we have read. They offer a (necessarily limited) panorama of a few applications of classical mechanics to modern physics. All problems are independent of each other; however, we have stressed in the solutions the links that exist between problems dealing with similar topics.

We have emphasized concrete applications (i.e. calculations of numerical values, estimates of orders of magnitude, interpretation of experimental data) in order to insist on the experimental character of physics, which is often not fully appreciated by students. In the same vein, we have given a number of reading suggestions at the end of the solutions, especially recent research articles (most of which can be found freely on the internet, especially on the arXiv preprint server). We believe that this feature of the book will reinforce in some of our readers their taste for physics.

It is probable that a few misprints remain in the text. Readers noticing them are invited to point them out to us by sending an e-mail to the address dgo@irsamc.ups-tlse.fr; we thank them in advance for their help.

David Guéry-Odelin
Thierry Lahaye

Acknowledgments

The preparation of this book benefited from countless interactions with many friends and colleagues.

First, we wish to thank Jean Dalibard for many inspiring discussions over the years, and for having taken the time to write the preface of this book.

We are also particularly indebted to Emmanuelle Archambault, Pascal Archambault, Gérald Bastard, Claude Cohen-Tannoudji, Antoine Couvert, Pierre Desbiolles, Laurent Gallot, Tomasz Kawalec, Tobias Koch, Thierry Lehner, Gonzalo Muga, Hélène Perrin, Tilman Pfau, Gaël Reinaudi, and Armin Ridinger, who sometimes directly inspired some of the problems presented here.

Last but not least, we would like to thank Valérie and Irina for their support.

Contents

Notations

We summarize below the main mathematical notations used in this book (we limit ourselves to notations that are often encountered). Specific notations are defined in the text when needed.

Symbol	Meaning
$\boldsymbol{a}$	Vector
$\boldsymbol{u}_x$	Unit vector along the direction x
$\boldsymbol{a} \cdot \boldsymbol{b}$	Dot product of the vectors $\boldsymbol{a}$ and $\boldsymbol{b}$
$\boldsymbol{a} \times \boldsymbol{b}$	Cross product of the vectors $\boldsymbol{a}$ and $\boldsymbol{b}$
M	Matrix
Tr(M)	Trace of the matrix M
Det(M)	Determinant of the matrix M
$\boldsymbol{\nabla}$	Gradient operator
$\langle f \rangle$	Time average of f
$\dot{x}$	Time derivative $\dfrac{\mathrm{d}x}{\mathrm{d}t}$
$\ddot{x}$	Second-order time derivative $\dfrac{\mathrm{d}^2 x}{\mathrm{d}t^2}$
∂_x	Partial derivative $\dfrac{\partial}{\partial x}$

Chapter 1

Orders of Magnitude and Dimensional Analysis

1.1 A few orders of magnitude *

When one solves a physics problem, it is important to rely on intuition, for instance to be able to guess the parameters on which an unknown quantity depends, or to estimate, before any calculation, the order of magnitude of the numerical value of the result. For that, it is necessary to know the orders of magnitudes of physical quantities that are characteristic of typical systems. This allows in particular to detect immediately obvious errors when one calculates a numerical value.

In the following exercise, one thus asks, for the most common physical quantities (lengths, masses, times, velocities. . .), the (approximate) numerical values corresponding to a few representative systems.

Universal constants

(1) What is the numerical value of the gravitational constant G?
(2) What is the numerical value of Coulomb's constant $1/4\pi\varepsilon_0$?
(3) What is the numerical value of the permeability of vacuum μ_0?
(4) What is the numerical value of Planck's constant h?
(5) What is the numerical value of Boltzmann's constant k_{B}?
(6) What is the numerical value of the ideal gas constant R?
(7) What is the numerical value of Avogadro's number $\mathcal{N}_{\mathrm{A}}$?

Lengths

(1) What is the typical size of an atomic nucleus?
(2) What is the typical size of an atom?
(3) To which range of wavelengths does the visible spectrum correspond?
(4) What is the value of the Earth's radius?

(5) What is the distance between the Earth and the Moon?

(6) What is the distance between the Earth and the Sun?

Masses

(1) What is the mass of an electron?

(2) What is the mass of a proton? And that of a neutron?

(3) What is the mass of the Earth?

(4) What is the mass of the Sun?

Densities

(1) What is the density of water?

(2) What is order of magnitude of the density of a usual metal?

(3) What is the density of air at the standard conditions for temperature and pressure?

Times

(1) What is the duration of a day in seconds?

(2) What is the duration of a year in seconds?

Frequencies

(1) What is the frequency range of electromagnetic waves corresponding to visible light?

(2) What is the frequency of electromagnetic waves used in a microwave oven?

(3) What is the frequency range used in radio broadcasting?

(4) To which frequency range do sound waves belong?

Velocities

(1) What is the typical speed of a satellite orbiting on a low-lying orbit around the Earth?

(2) What is the speed of sound in air?

(3) What is the speed of light in vacuum?

Energies

(1) What is the typical order of magnitude of a molecular binding energy?

(2) What is the typical energy of an atom in an ideal monoatomic gas at temperature T?

(3) What is the energy required to increase the temperature of one gram of water by one degree Celsius?

Powers

(1) What is the typical power of a light bulb?
(2) What is the typical power of a car engine?
(3) What is the typical power of a nuclear power plant? And that of a wind turbine?
(4) What is the average power consumed by the human body?

Solution

Universal constants

(1) The gravitational constant G is about $6.67 \times 10^{-11}\,\mathrm{m}^3 \cdot \mathrm{kg}^{-1} \cdot \mathrm{s}^{-2}$.
(2) $1/4\pi\varepsilon_0 \simeq 9 \times 10^9\,\mathrm{m} \cdot \mathrm{F}^{-1}$.
(3) By definition $\mu_0 = 4\pi \times 10^{-7}\,\mathrm{H} \cdot \mathrm{m}^{-1}$. Note that one has the *exact* relationship $\mu_0\varepsilon_0 c^2 = 1$. Since μ_0 and c have an *exact* value, so has ε_0.
(4) Planck's constant is approximately $6.63 \times 10^{-34}\,\mathrm{J} \cdot \mathrm{s}$. One often uses the *reduced Planck's constant* $\hbar \equiv h/2\pi \simeq 1.05 \times 10^{-34}\,\mathrm{J} \cdot \mathrm{s}$.
(5) Boltzmann's constant k_{B} is approximately equal to $1.38 \times 10^{-23}\,\mathrm{J} \cdot \mathrm{K}^{-1}$.
(6) The ideal gas constant has the numerical value $R \simeq 8.314\,\mathrm{J} \cdot \mathrm{K}^{-1} \cdot \mathrm{mol}^{-1}$.
(7) One has $\mathcal{N}_{\mathrm{A}} \simeq 6.022 \times 10^{23}\,\mathrm{mol}^{-1}$. Note that the ideal gas constant is related to Boltzmann's constant and to Avogadro's number by $R = \mathcal{N}_{\mathrm{A}} k_{\mathrm{B}}$.

Lengths

(1) The typical size of an atomic nucleus is on the order of one femtometer ($1\,\mathrm{fm} = 10^{-15}$ m).
(2) The typical size of an atom is on the order of 10^{-10} m, i.e. an Angstrom (Å). More precisely, the Bohr radius of hydrogen is $a_0 = 0.53$ Å.
(3) The visible spectrum extends from the violet to the red, i.e. from 400 to 800 nm.
(4) The Earth's radius is about 6,400 km, implying a circumference of 40,000 km (the first definition of the meter, dating back to the French Revolution, defined the meter as "the ten-millionth part of one quarter of the Earth meridian".)

(5) The distance between the Earth and the Moon is about 384,000 km, i.e. a little bit more than a light-second.

(6) The distance between the Earth and the Sun (which defines the *astronomical unit*) is about 150 million kilometers (or 8 light-minutes).

Masses

(1) The mass of an electron is $m_e \simeq 9.1 \times 10^{-31}$ kg. One can also remember its value expressed in energy units (*via* the relationship $E = mc^2$), which is 511 keV (we recall that $1\,eV = 1.6 \times 10^{-19}$ J).

(2) The neutron and the proton have almost the same mass, on the order of 1.67×10^{-27} kg, or 938 MeV. The proton is about 1,836 times as massive as the electron. The neutron is in fact very slightly more massive than the proton, which makes the β decay of a free neutron (into a proton, an electron and an antineutrino) energetically possible.

(3) The mass of the Earth is about 6×10^{24} kg.

(4) The mass of the Sun is about 2×10^{30} kg.

Densities

(1) The density of water is $1{,}000$ kg $\cdot$ m^{-3}.

(2) The density of iron, for example, is 7,900 kg$\cdot$m^{-3}; the density of copper 8,900 kg $\cdot$ m^{-3}. Some metals, like the alkalis, have a very small density (like lithium, 530 kg $\cdot$ m^{-3}). The density of mercury is 13,530 kg $\cdot$ m^{-3}; the density of platinum (used, together with iridium, for the realization of the kilogram prototype) is 21,450 kg $\cdot$ m^{-3}.

(3) The density of air at the standard conditions for temperature and pressure (273 K and 10^5 Pa) is

$$\rho = \frac{PM}{RT} \simeq 1.3 \text{ kg} \cdot \text{m}^{-3}.$$

Times

(1) A *solar* day (24 hours) has a duration of $24 \times 3{,}600 = 86{,}400$ s. The *sidereal* day has a duration of 86,164 s.

(2) A year is 365.25 days, or about 30 million seconds.

Frequencies

(1) The visible domain corresponds to frequencies ranging from 4×10^{14} Hz (red light) to 8×10^{14} Hz (violet light).

(2) A microwave oven uses microwaves at 2.45 GHz.

(3) For radio broadcasting, one uses waves with frequencies around 100 kHz (for amplitude modulation) or 100 MHz (for frequency modulation).
(4) The human ear is sensitive to sound waves of frequencies ranging from 20 Hz to 20 kHz.

Velocities

(1) A satellite on a low orbit around the Earth has a velocity on the order of $\sqrt{GM_T/R_T}$, with M_T the mass of the Earth and R_T its radius. One finds a speed of about 8 km·s^{-1}. It thus takes about $40,000/8 \simeq 5,000$ s (a little bit more than one hour) for the satellite to go round the Earth.
(2) Sound waves need a medium to propagate. The speed of sound in air, at ambient temperature, is about 340 m·s^{-1}. For a gas of molar mass M, at temperature T, one has $c_{\text{sound}} = \sqrt{\gamma RT/M}$, where R is the ideal gas constant and γ the ratio of specific heats at constant pressure and at constant volume ($\gamma = 1.4$ for a gas of diatomic molecules such as air). In water, the speed of sound is about 1.5 km·s^{-1}.
(3) The speed of light in a vacuum has (by definition of the meter) the fixed value 299,792,458 m·s^{-1}. One uses in general $c \simeq 3 \times 10^8$ m·s^{-1}. In a transparent medium with refractive index n, the velocity of light is c/n.

Energies

(1) Molecular binding energies are on the order of a few electronvolts ($1\,\text{eV} = 1.6 \times 10^{-19}$ J).
(2) The energy of an atom in a monoatomic ideal gas at temperature T is $3k_B T/2$, i.e. for $T = 300$ K, on the order of 6×10^{-21} J.
(3) The energy required to increase, by one degree Celsius, the temperature of one gram of water is what defines the *calorie*, and is equivalent to 4.18 J.

Powers

(1) A light bulb has a typical power of a few tens of watts.
(2) Even today, one measures the power of car engines in *horsepower*, which corresponds to 735 W. A 100 hp car therefore has a power of 74 kW.
(3) A nuclear power plant has a power ranging from 100 to 1,000 MW. A wind turbine has a typical power of 1 MW.
(4) It is well known that a human being must consume about 2,000 kilocalories per day. As one kilocalorie corresponds to 4.18 kJ, the average

power dissipated by a human being is

$$P \simeq \frac{2 \times 10^3 \times 4.18 \times 10^3}{24 \times 3,600} \simeq 100 \text{ W}.$$

The instantaneous power can reach several kW, e.g. when performing physical effort.

1.2   Dimensional analysis *

In this exercise, one applies, to a few simple examples, the technique of dimensional analysis, which allows one to obtain, without any calculation, the form of the solution of a physics problem, by combining the relevant physical parameters of the problem into quantities that have the same physical dimension (length, mass, time...) as the results[1]. In cases where the number of parameters is small, one often encounters the situation in which only one combination of the parameters has the right dimension. One thus obtains the correct result within a (dimensionless) numerical factor, which is often close to unity. For numerical calculations, one will use the values recalled in the solution of Problem 1.1.

Free fall.   A particle of mass m falls from a height ℓ, without initial velocity, in the gravity field g. How does the fall time scale with ℓ and g?

Classical radius of the electron.   One assumes that the electron (mass m_e, charge q_e) can be described by a spherical charge distribution, with a radius r_e. In the framework of classical (i.e. non-quantum) relativistic theory, calculate r_e. What is the numerical value of r_e? Comment.

Bohr radius of the hydrogen atom.   Can one find, by dimensional analysis, the typical size of the orbit of an electron around a proton, in the framework of classical mechanics? Why? Show that introducing, in quantum mechanics, the reduced Planck's constant $\hbar$ allows one to obtain the characteristic size of the orbit (hint: $\hbar$ has the dimension of an *action*, i.e. of an energy multiplied by a time, or, equivalently, of an angular momentum).

---

[1]It is possible to make dimensional analysis formal by using the so-called "π-theorem": see e.g. E. Buckingham, *On physically similar systems: Illustrations of the use of dimensional analysis*, Phys. Rev. **4**, 345 (1914).

$t = 4$ s $\qquad t = 8$ s $\qquad t = 16$ s $\qquad t = 28$ s $\qquad t = 46$ s

Fig. 1.1 *Explosion of an American nuclear bomb in 1953. The time indicated above each image corresponds to the time elapsed since the explosion. The scale is the same for all photographs:* 4 km × 2.5 km.

Plasma frequency. One considers a gas of electrons (with number density n) moving in a background of static ions, in such a way that the system is neutral. Find by dimensional analysis the expression of the characteristic oscillation frequency of this plasma.

Stokes' formula. Find by dimensional analysis the form of the drag force exerted by a fluid with viscosity η and density ρ on a sphere of radius R, moving with velocity v in this fluid. Hint: the *coefficient of (dynamical) viscosity* η has the dimension of a pressure divided by a (spatial) gradient of velocity. Is there *a priori* only one solution? To go further, one will assume that the force is proportional to the velocity.

Energy released in a nuclear explosion. In the fifties, the energy released in a nuclear explosion was classified information. However, movies showing the expansion of the famous "nuclear mushroom" were released to the public by the US Army. The British physicist G.I. Taylor then deduced the energy released by the explosion, by assuming that the radius R of the cloud depended only on the time t elapsed since the explosion, on the released energy E, and on the density ϱ of the air around the cloud. Show that, under those assumptions, $R \sim t^\beta$. What is the value of β? Show that by measuring $R(t)$ one can deduce E. Figure 1.1 presents a series of photographs of a thermonuclear explosion, taken for various t. Is the law $R(t)$ obtained by dimensional analysis in agreement with this series of images?

Solution

In all that follows, one denotes by $[A]$ the dimension of the physical quantity A. We recall that the three fundamental dimensions in mechanics are the length, noted L, the mass, noted M, and the time, noted T. Thus the equation $[v] = L \cdot T^{-1}$ means that the velocity v has the dimension of a length divided by a time.

Free fall.   One looks for an expression of the form $t \sim m^\alpha g^\beta \ell^\gamma$ for the free fall time, which implies, dimensionally:

$$T = M^\alpha (L \cdot T^{-2})^\beta L^\gamma$$

or, identifying the various exponents of M, L and T, $\alpha = 0$, $\beta + \gamma = 0$, and $-2\beta = 1$. This set of equations has the solution $(\alpha, \beta, \gamma) = (0, -1/2, 1/2)$, and the free fall time thus scales as $t \sim \sqrt{\ell/g}$, which is the well-known result, within a factor $\sqrt{2}$.

Classical radius of the electron.   In the framework of classical electrodynamics, the only parameters that can enter the expression of r_e are the electron charge q_e, or in fact the combination $q_e^2/(4\pi\varepsilon_0)$, which has the dimension of an energy multiplied by a length (ML^3T^{-2}), the electron mass m_e, and the speed of light c. One finds

$$r_e = \frac{q_e^2}{4\pi\varepsilon_0 m_e c^2} \simeq 2.82 \times 10^{-15}\,\text{m}.$$

The classical radius of the electron thus has the same order of magnitude as the size of the atomic nucleus; however, contrary to nuclei, electrons are, as far as is known today, *point-like elementary particles*. The classical radius of the electron only gives the typical size an electron would have *if* one could describe it classically. Quantum mechanics substitutes to this notion of radius the concept of a probability density over a distance deduced from quantum electrodynamics.

Bohr radius of the hydrogen atom.   In the framework of classical, non-relativistic mechanics, the only parameters that can play a role in the expression of the orbit radius are $q_e^2/(4\pi\varepsilon_0) \equiv e^2$ (see previous question), and the electron mass m_e. Since $[e^2] = ML^3T^{-2}$ and $[m_e] = M$, we have no way of eliminating the time appearing in $[e^2]$. One thus cannot, simply by dimensional analysis, obtain the radius of the electron orbit in the hydrogen atom. In fact, one expects this result, since it is well known for the Kepler problem in the case of gravitational interactions (having the same $1/r$ dependence as the Coulomb interaction) that the orbits of the planets around the Sun, for instance, have an arbitrary size, fixed by initial conditions. In that case, there is no characteristic length scale, and the same laws explain the motion of satellites around the Earth, of planets around the Sun and of the Sun around the galactic center[2].

In quantum mechanics, Planck's constant $\hbar$, which is an *action* with dimension ML^2T^{-1} (linear momentum $\times$ length, or, equivalently, energy $\times$ time), allows one to fix the size a_0 of the hydrogen atom. Indeed, by writing

$$a_0 = (e^2)^\alpha \hbar^\beta m_e^\gamma$$

one must have, in order to obtain a length,

$$\begin{cases} \alpha & + \beta & + \gamma = 0 \\ 3\alpha & + 2\beta & = 1 \\ -2\alpha & - \beta & = 0 \end{cases}$$

with solution $(\alpha, \beta, \gamma) = (-1, 2, -1)$. One thus finds, for the expression of the size of the electron orbit in the hydrogen atom, the following value:

$$a_0 = \frac{\hbar^2}{m_e e^2} = \frac{4\pi\varepsilon_0 \hbar^2}{m_e q_e^2}.$$

This quantity is called the *Bohr radius*, after the Danish physicist Niels Bohr, who devised the first quantum-mechanical model (in fact, a semiclassical one) of the atom. Numerically, one has $a_0 \simeq 0.53 \times 10^{-10}$ m.

♦ **Remark.** Using e^2, $\hbar$ and c one can build a dimensionless number, the *fine structure constant*, denoted α, equal to

$$\alpha \equiv \frac{e^2}{\hbar c} \simeq \frac{1}{137.04}.$$

It measures the strength of electromagnetic interactions. Note that the Bohr radius can be expressed, as a function of the classical radius of the electron, as:

$$a_0 = \frac{r_e}{\alpha^2}.$$

Plasma frequency. This frequency can only depend on q_e^2/ε_0, on the density n, and on the electron mass m_e. One thus searches an expression of the form:

$$\omega_p = n^\alpha \left(\frac{q_e^2}{\varepsilon_0}\right)^\beta m_e^\gamma.$$

---

[2]Let us stress, however, a fundamental difference between gravitational and Coulomb interactions: in the latter case, the existence of positive *and* negative charges implies that the long-range character of the interaction is often "masked" by screening of a charge by other charges of opposite sign. On the contrary, mass being always positive, the gravitational interaction is never screened; this is why it plays a crucial role on astronomical scales, although its strength is very weak compared to that of Coulomb interaction.

This implies, dimensionally:

$$T^{-1} = L^{-3\alpha} \left(ML^3T^{-2}\right)^\beta M^\gamma,$$

which leads to

$$\begin{cases} -3\alpha + 3\beta & = 0 \\ \beta + \gamma = 0 \\ -2\beta & = -1 \end{cases}$$

with solution $(\alpha, \beta, \gamma) = (1/2, 1/2, -1/2)$. One thus obtains:

$$\omega_{\text{p}}^2 = \frac{nq_{\text{e}}^2}{m_{\text{e}}\varepsilon_0}.$$

Stokes' formula. Let us note that if one writes an expression of the form

$$F \sim \eta^\alpha \rho^\beta v^\gamma R^\delta,$$

dimensional analysis will give only three equations for four unknowns. One thus needs to use the extra assumption given in the text, i.e. the force is proportional to the velocity ($\gamma = 1$). One gets

$$MLT^{-2} = \left(ML^{-1}T^{-1}\right)^\alpha \left(ML^{-3}\right)^\beta LT^{-1}L^\delta,$$

from which one deduces the set of equations:

$$\begin{cases} \alpha + \beta & = 1 \\ -\alpha - 3\beta + \delta = 0 \\ -\alpha & = -1 \end{cases}$$

whose solution reads $(\alpha, \beta, \delta) = (1, 0, 1)$, and thus we have the following scaling

$$F \sim \eta R v.$$

Stokes' formula actually reads $F = 6\pi\eta Rv$.

♦ **Remark.** The fact that there is not a unique solution if one does not fix the exponent γ comes from the fact that it is possible to build, using the four parameters η, ρ, R, and v, a dimensionless number:

$$\text{Re} \equiv \frac{Rv\rho}{\eta}.$$

This number is called the *Reynolds number* and allows one to determine whether a flow is dominated by viscosity (small Reynolds number Re $\ll 1$), or by inertia (Re $\gg 1$). In the latter case, the flow is turbulent. As far as the drag force on a sphere is concerned, it can be written in the most general case $F = \eta Rvf(\text{Re})$, where f is a dimensionless function. Stokes' formula implies that $\lim_{x \to 0} f(x) = 6\pi$. For large Reynolds numbers, one can expect that the viscosity does not appear any more in the expression of the drag force. By dimensional analysis, one then finds that $F \sim \rho R^2 v^2$: the drag is proportional to the square of the velocity and to the cross-section of the sphere. The function f thus fulfills, when $x \to \infty$, $f(x) \sim Ax$, where A is a constant.

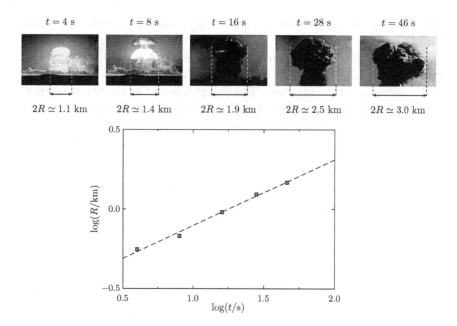

Fig. 1.2 *Above, we have indicated for each image the extension $2R(t)$ of the nuclear "mushroom". Bottom: plot of $\log(R/\text{km})$ versus $\log(t/\text{s})$. One finds a straight line with slope 0.41 (dashed line), very close to the value $2/5 = 0.4$ obtained by dimensional analysis. The intercept is -0.52.*

Energy released in a nuclear explosion. Let us write an equation of the form

$$R \sim \varrho^{\alpha} t^{\beta} E^{\gamma}.$$

Dimensional homogeneity implies:

$$L = \left(ML^{-3}\right)^{\alpha} T^{\beta} \left(ML^2 T^{-2}\right)^{\gamma},$$

giving the following system:

$$\begin{cases} -3\alpha & + 2\gamma = 1 \\ \alpha & + \gamma = 0 \\ \beta & - 2\gamma = 0 \end{cases}$$

which has the solution $(\alpha, \beta, \gamma) = (-1/5, 2/5, 1/5)$. One thus deduces

$$R \sim \frac{E^{1/5} t^{2/5}}{\varrho^{1/5}}.$$

On the series of photographs, one measures the size R of the nuclear "mushroom" (for instance its horizontal diameter $2R$). By plotting $\log(R)$

versus $\log(t)$, one indeed finds a straight line, with a slope close to 2/5 (see Fig. 1.2). To obtain the energy E released by the explosion, one determines the intercept $\log(E/\varrho)/5 = -0.52$ of this line. With $\varrho \simeq 1.3$ kg $\cdot$ m^{-3}, one finds $E \sim 3 \times 10^{12}$ J (or about one kiloton of TNT, to use the "traditional" unit associated with explosives). The energy released by the bomb dropped onto Hiroshima on August 6th, 1945[3] was 15 kilotons of TNT.

---

[3] A bomb which caused more than 70,000 casualties...

Chapter 2

Motion in the Gravity Field

2.1 Safety parabola *

A projectile is fired from a point O in the Earth's gravity field, with an initial velocity v_0 whose modulus is fixed, but which has an arbitrary direction. Determine the *safe area*, i.e. the positions in which one cannot be reached by the projectile, regardless of the angle α that the initial velocity makes with the horizontal plane.

> **Solution**

The initial velocity v_0 and the acceleration due to gravity g define a vertical plane. We denote by (O, x, y) the orthonormal frame associated with this plane, with y as the ascending vertical axis: $g = -g u_y$ (u_y is the unit vector along the y axis). The equations of motion for the projectile read $\ddot{x} = 0$ and $\ddot{y} = -g$. Their integration is straightforward and, taking into account the initial conditions, yields:

$$\begin{cases} x(t) = v_0 \cos \alpha \, t \,, \\ y(t) = -gt^2/2 + v_0 \sin \alpha \, t \,. \end{cases}$$

By eliminating between these two equations the explicit time variable t, one obtains the equation for the trajectory, which is a parabola:

$$y = \tan \alpha \, x - \frac{g}{2v_0^2} \frac{x^2}{\cos^2 \alpha} \,. \tag{2.1}$$

Let us search for the condition that the angle α has to fulfill for a given point of coordinates (x, y) to belong to the family of parabolæ (2.1). Using the relation $1/\cos^2 = 1 + \tan^2$, we get:

$$\frac{gx^2}{2v_0^2} \tan^2 \alpha - x \tan \alpha + \frac{gx^2}{2v_0^2} + y = 0 \,,$$

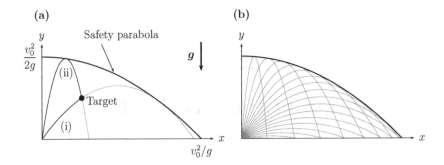

(a) **(b)**

Fig. 2.1 (a) *There are two different possible trajectories to reach a point located below the safety parabola.* (b) *The safety parabola is the envelope of the family of parabolic trajectories obtained for all possible values of the angle α of the initial velocity with respect to the horizontal axis x.*

i.e. $\tan \alpha$ is the solution of a quadratic equation. The point (x, y) is in the so-called "safe domain" if and only if there is no solution of the quadratic equation whatever the value of α, or otherwise stated, if and only if the discriminant of the quadratic equation is negative:

$$y > \frac{v_0^2}{2g} - \frac{g}{2v_0^2} x^2 \,. \tag{2.2}$$

The points in the "safe area" are therefore located above a parabola (see Fig. 2.1). If a point of coordinate (x, y) is below this parabola, there exist two real solutions for $\tan \alpha$, and thus two possible angles α for the point to be reached. We find here a result which is well known by basketball players: there are two trajectories to reach a target, by a straight shot (i) or by a bell-like trajectory (ii) [see Fig. 2.1(a)].

Let us describe another approach to derive the equation for the safety parabola. In the field of gravity, each trajectory is a parabola whatever the angle α of the initial velocity of the projectile with respect to the horizontal axis x. If α is varied, those parabolæ (2.1) form a family of trajectories. The safety parabola is nothing but the envelope of all those trajectories, and turns out to be another parabola that separates the points which can be shot from those which are out of reach. To find the envelope of the family of parabolæ $F(x, y, \alpha) = y - \tan \alpha \, x + gx^2/(2v_0^2 \cos^2 \alpha)$, one needs to solve the following set of equations (see reminder below):

$$\begin{cases} F(x, y, \alpha) = 0 \,, \\ \dfrac{\partial F(x, y, \alpha)}{\partial \alpha} = 0 \,. \end{cases}$$

One readily shows that this set of two equations is equivalent to Eq. (2.2). Figure 2.1(b) represents the family of trajectories obtained by varying the angle α, and their envelope.

▶ **Further reading.** The article by J.-M. Richard, *Safe domain and elementary geometry*, Eur. J. Phys. **25**, 835 (2004), also freely available on the internet at the address http://arxiv.org/abs/physics/0410034, studies by geometrical methods the safe domains for the case of several force fields, such as Coulomb and harmonic potentials.

Reminder: Envelope of a family of curves

We consider, in the plane (Oxy), a family of curves $\mathcal{C}_\lambda$ that are defined by the equation $F(x, y; \lambda) = 0$ depending on the parameter λ. When λ varies by a small quantity $d\lambda$, the curve $\mathcal{C}_\lambda$ becomes $\mathcal{C}_{\lambda+d\lambda}$. A point (x, y) belongs to both curves if and only if $F(x, y; \lambda) = F(x, y; \lambda + d\lambda) = 0$, or, equivalently,

$$\begin{cases} F(x, y; \lambda) = 0, \\ \dfrac{\partial F(x, y; \lambda)}{\partial \lambda} = 0. \end{cases} \qquad (2.3)$$

Such a point is called a *characteristic point* of the family of curves. Equations (2.3) define a curve which is the locus of characteristic points, and which is called the *envelope* of the family of curves $\mathcal{C}_\lambda$. In geometrical optics, the envelope of a family of rays of light is called a *caustic*.

2.2 Shape of a liquid jet *

Figure 2.2 shows water flowing out of a funnel. One observes that the jet radius decreases when the fall height z increases[1].

(1) Explain qualitatively the origin of this effect.
(2) Calculate $R(z)$ (the origin of the z-axis is taken at the opening of the funnel).
(3) By analyzing the photograph in Fig. 2.2, one obtains a table of values $R(z)$. Are the experimental results in agreement with the theoretical prediction? Hint: use a graphical method.

---

[1]The reader is invited to perform the experiment by him/herself (in his/her bathroom...): one just needs a plastic bottle, in which one drills a small hole (one to two millimeters in diameter) at the bottom, to fill it up, turn it upside down, and let the water flow through the bottleneck. The small hole at the bottom allows air (which replaces water as it flows out of the bottle) to enter the bottle not through the bottleneck, otherwise the flow would not be laminar.

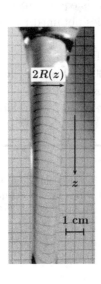

z/mm	$R(z)$/mm
0	18.3
5	15.1
10	13.1
15	12.3
20	11.4
25	11.1
30	10.3
35	10.0
40	9.6
45	9.3
50	9.1
55	8.9
60	8.8
65	8.6
70	8.6
75	8.3
80	8.1
85	7.9

Fig. 2.2 *A water jet flows under the influence of gravity. The radius R of the jet decreases when the fall height z increases.*

Solution

(1) Because of the acceleration due to gravity, the vertical velocity $v(z)$ of the fluid increases with increasing z. Since water is incompressible, the conservation of mass implies the conservation of volumetric flow rate Q. As Q is equal to the product of the velocity by the cross section of the jet, one has

$$v(z)\, R^2(z) = \text{const} = v_0\, R_0^2, \qquad (2.4)$$

where v_0 is the velocity of the flow at $z = 0$, and R_0 the jet radius at $z = 0$. This explains why $R(z)$ decreases when z increases.

(2) In order to find $R(z)$, it is sufficient to know $v(z)$. One then has to solve an elementary problem in fluid dynamics: by applying Bernoulli's principle (see reminder below), one obtains the so-called Torricelli formula:

$$v^2(z) = v_0^2 + 2gz. \qquad (2.5)$$

The liquid velocity is the same as the one of a free-falling particle.

Combining Eqs (2.4) and (2.5) one finds the jet radius as a function of z:

$$R(z) = \frac{R_0}{(1 + 2gz/v_0^2)^{1/4}}. \tag{2.6}$$

(3) In order to test Eq. (2.6), a first verification consists in checking that for large z, one has $R \sim 1/z^{1/4}$. For that, one plots $\log R$ versus $\log z$ (see Fig. 2.3). One gets a curve whose asymptote is indeed a straight line of slope $-1/4$, that corroborates the scaling law (2.6).

To go further and test Eq. (2.6) more quantitatively, one can plot $[R_0/R(z)]^4$ as a function of z. One indeed gets a straight line with intercept 1 (see Fig. 2.4), and whose slope is $2g/v_0^2 \simeq 0.31$ mm^{-1}, which allows one to determine the velocity at $z = 0$:

$$v_0 \simeq 25\,\text{cm} \cdot \text{s}^{-1}.$$

Reminder: Bernoulli's principle

For an ideal (no viscosity and no thermal conductivity), stationary and incompressible flow, Bernoulli's principle states that

$$P/\rho + gz + v^2/2 = \text{const}$$

along a streamline (P is the pressure, ρ the fluid density, and v its velocity). Note that for $v = 0$, one recovers, as expected, the laws of hydrostatics for an incompressible fluid.

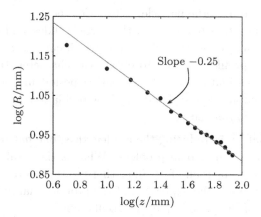

Fig. 2.3 *Radius $R(z)$ of the jet as a function of z, in a log-log plot.*

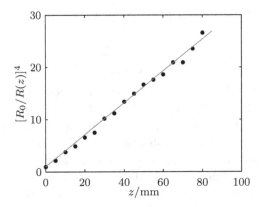

Fig. 2.4   *Graphical determination of the initial velocity of water at the output of the funnel.*

2.3   The tautochrone curve **⋆⋆**

In order to illustrate experimentally the behavior of a one-dimensional harmonic oscillator, i.e. of a particle moving along say Ox and subjected to a quadratic trapping potential, one can think of making a particle slip, in the gravity field $\boldsymbol{g} = -g\boldsymbol{u}_z$, on a slide of equation $z(x) = x^2/\ell$ (where ℓ is a length, and Oz the vertical direction).

(1) Show that, strictly speaking, such a configuration does not lead to a harmonic motion of the particle. Why? Comment.
(2) Why is it important to state that the particle slides, and does not roll?
(3) In the following, we want to determine which curve $z(x)$ must be used for the slide profile, in order to obtain the characteristic behavior of a harmonic oscillator, i.e. an oscillation period independent of the oscillation amplitude (*isochronism* of the oscillations). Such a curve is called a *tautochrone* curve[2].

 (a) Show that by introducing the arc length s, one has to solve an explicitly one-dimensional problem. What is the total energy of the particle as a function of s? Show that the tautochrone curve fulfills $z(s) \propto s^2$. Give the expression of the proportionality constant as a function of the period T_0 of the oscillations.
 (b) Deduce from the above the parametric equations of the tautochrone

---

[2]This word is formed on the Greek radicals ταυτὸ, meaning "the same", and χρόνος, for "time".

curve. One may use the angle θ such as $\tan\theta = z'(x)$. What is the geometrical nature of this curve?

Solution

(1) The potential energy in the gravity field reads $E_{\mathrm{p}}(x) = mgz = mgx^2/\ell$. One finds an energy which is quadratic in x. This seems to imply, at first sight, that the motion will be harmonic along Ox. However this is not the case. Indeed, the kinetic energy reads $E_{\mathrm{k}} = mv^2/2 = m(\dot{x}^2 + \dot{z}^2)/2$. We thus have the extra term $m\dot{z}^2/2$, and the total energy does not read simply

$$E = m\dot{x}^2/2 + m\omega^2 x^2/2,$$

where $\omega^2 = 2g/\ell$. Noticing that $\dot{z}/\dot{x} = (\mathrm{d}z/\mathrm{d}t)/(\mathrm{d}x/\mathrm{d}t) = \mathrm{d}z/\mathrm{d}x = z'(x)$ (since the trajectory of the particle is $z(x)$), we obtain for the expression of the total energy:

$$\begin{aligned}
E &= m\frac{\dot{x}^2}{2}\left(1 + z'(x)^2\right) + \frac{1}{2}m\omega^2 x^2 \\
&= m\frac{\dot{x}^2}{2}\left(1 + \frac{4x^2}{\ell^2}\right) + \frac{1}{2}m\omega^2 x^2,
\end{aligned}$$

which proves that the motion will be approximately harmonic only in the limit $x \ll \ell$ (small oscillations).

(2) If a ball of radius R *rolls without sliding*, the kinetic energy includes, in addition to the kinetic energy of the center of mass, a term corresponding to the spin of the ball, equal to $J\Omega^2/2$, where $J = 2mR^2/5$ is the moment of inertia of the ball, and Ω its rotation speed, which fulfills $\Omega = \sqrt{\dot{x}^2 + \dot{z}^2}/R$ (condition of rolling without sliding). Therefore the dynamics of a rolling ball is different from that of a sliding particle (the ball is slower).

(3) (a) Introducing the arc length s, the velocity (in magnitude) simply reads $\dot{s}$. We infer the expression of the total energy of the particle:

$$E = \frac{1}{2}m\dot{s}^2 + mgz.$$

Let us try to obtain an expression depending only on the variable s, in order to have a one-dimensional problem. To have the expression of the energy of a harmonic oscillator (which ensures the isochronism of the oscillations), we must have:

$$z(s) = \frac{s^2}{\ell}, \tag{2.7}$$

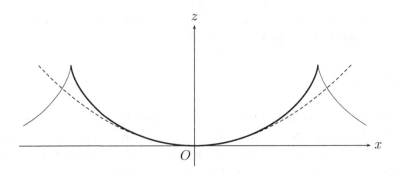

Fig. 2.5   *The tautochrone curve (solid line), together with its osculating parabola (dashed line). Obviously, only one arch (thick line) actually corresponds to the tautochrone curve.*

where the constant ℓ, homogeneous to a length, is such that the angular frequency ω of the oscillations fulfills $\omega^2 = 2g/\ell$. Expressed as a function of the oscillation period $T_0 = 2\pi/\omega$, the length ℓ thus reads:

$$\ell = \frac{gT_0^2}{2\pi^2}.$$

(b) Let us first determine the function $z(\theta)$. By differentiating Eq. (2.7) and squaring, we get:

$$(dz)^2 = \frac{4z}{\ell}(ds)^2.$$

Now, by definition of the arc length, one has

$$(ds)^2 = (dx)^2 + (dz)^2 = (dz)^2\left(1 + \frac{1}{\tan^2\theta}\right) = \frac{(dz)^2}{\sin^2\theta},$$

where one has used $dz/dx = \tan\theta$. Therefore, combining the last two equations, one finds:

$$z = \frac{\ell}{4}\sin^2\theta. \tag{2.8}$$

To determine $x(\theta)$, one starts from

$$dx = \frac{dz}{\tan\theta} = \frac{\ell}{2}\cos^2\theta\,d\theta$$

(where, in the second equality, we have used the explicit expression (2.8) of z as a function of θ), which gives upon integration

$$x = \frac{\ell}{8}\left[2\theta + \sin(2\theta) + C\right]$$

where C is a constant that can be chosen equal to zero (this amounts to choosing the origin at the minimum of the curve $z(x)$). Finally, one obtains the parametric equations of the tautochrone curve:

$$\begin{cases} x(\varphi) = (\ell/8)(\varphi + \sin\varphi), \\ z(\varphi) = (\ell/8)(1 - \cos\varphi), \end{cases}$$

where $\varphi = 2\theta$. Note that one recovers, close to the origin ($\varphi \ll 1$), $z \simeq x^2/\ell$. These parametric equations are those of a cycloid (see reminder below), i.e. the curve traced by a point of a circle (here, of radius $\ell/8$) rolling without sliding along a straight line (here, the line of equation $z = \ell/4$). Figure 2.5 represents the corresponding curve.

♦ **Remark.** The cycloid possesses another remarkable property as far as motion in the gravity field is concerned. One can show that the profile one must use for a curve passing through two given points in order to minimize the transit time of a particle sliding between those two points under the influence of gravity (the so-called *brachistochrone* problem), is also a cycloid. This can be shown using the *calculus of variations*.

Reminder: Parametric equations of the cycloid

We consider a circle of center C and radius R which is rolling (without sliding) onto the horizontal line $y = 0$ — see figure below:

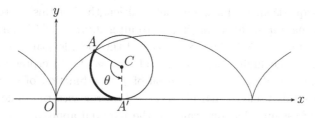

We wish to find the coordinates of the point A on the periphery of the circle, which is initially at the origin O of the coordinate system. For that purpose, we notice that when the circle has turned by an angle θ, the coordinates of its center C are $(R\theta, R)$ since the length of the arc OA' is $R\theta$ (no-sliding condition). By elementary trigonometry, the coordinates of A are easily found to be $(R\theta - R\sin\theta, R - R\cos\theta)$, which are the parametric equations of the cycloid.

Since the y-coordinate is 2π-periodic in θ, the cycloid is made up of an infinity of arches. We leave as an exercise to the reader to show that the length of an arch of cycloid is $8R$, and the area under it, $3\pi R^2$.

2.4   Gravitational cavity **

We consider atoms that are constrained to move in the vertical plane (Oxy), and that bounce on a surface of equation $y = ax^2$, which plays the role of a mirror (Fig. 2.6). The bounces are supposed to be perfectly elastic. The curvature of the mirror allows one to trap the atoms having an horizontal velocity component, which would be impossible without a curvature. The goal of this exercise is to characterize those trapped trajectories.

The atoms are only subjected to gravity $g = -gu_y$. In the following, we focus on a single atom, whose position is $(x_n, y_n = ax_n^2)$ and whose velocity is $(v_{x,n}, v_{y,n})$ just after the nth bounce.

(1) Show that the abscissa x_{n+1} of the $(n + 1)^{\text{th}}$ impact on the mirror fulfills:

$$x_{n+1} = \frac{1}{g + 2av_{x,n}^2} \left[2v_{x,n}v_{y,n} + x_n(g - 2av_{x,n}^2) \right].$$

Hint: one may denote by t^* the duration of the free flight between the nth and the $(n + 1)^{\text{th}}$ collisions.

(2) Let v_- be the velocity of the atom just before the $(n + 1)^{\text{th}}$ bounce; its components are written as (v_x, v_y). What is the value of v_x? Show that $v_{x,n+1}$ reads:

$$v_{x,n+1} = \frac{1}{1 + 4a^2x_{n+1}^2} \left[v_{x,n}(1 - 4a^2x_{n+1}^2) + 4ax_{n+1}v_y \right].$$

(3) In this question, we focus on the focal length f of this mirror. Find the ordinate y of the atom (in $x = 0$) after its $(n + 1)^{\text{th}}$ bounce in the case where $v_{x,n} = 0$. The expression of the focal length f is obtained by taking the limits $v_{y,n} \to \infty$ and $ax_n \ll 1$. Interpret these two conditions, and give the expression of f as a function of a. One will compare this value to that of an optical mirror with the same shape.

(4) We now assume that one can use the paraxial approximation, which corresponds to the assumptions: $ax_n \ll 1$, $av_{x,n}^2 \ll g$ and $v_{x,n} \ll v_{y,n}$, so that $v_{y,n} \simeq v_y$ is barely affected by the bounces of the atom. We denote by H the height reached by the atom after the nth bounce. Determine the expression of H as a function of v_y and g. Show that, under these assumptions:

$$\begin{pmatrix} x_{n+1} \\ v_{x,n+1} \end{pmatrix} = \begin{pmatrix} 1 & 2v_y/g \\ -4av_y & 1 - 8av_y^2/g \end{pmatrix} \begin{pmatrix} x_n \\ v_{x,n} \end{pmatrix}.$$

Deduce a criterion on H and f determining the stability of the trajectory of the atom for an arbitrary number of bounces.

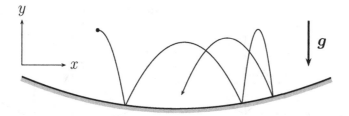

Fig. 2.6 *An atom bouncing on a parabolic mirror in the gravity field.*

Solution

Before starting to solve the exercise, let us calculate the position of the focus for an optical parabolic mirror. The surface of the mirror has the equation $y = ax^2$. In order to obtain the position of the focus F, we consider a light ray coming from infinity along the unit vector $-\boldsymbol{u}_y$ [see Fig. 2.7(a)]. It is reflected at $(x^\star, y^\star = ax^{\star 2})$. The normal $\mathcal{D}$ to the parabola at this point makes the angle α with the (Oy) axis, and one has $\tan \alpha = y'(x^\star) = 2ax^\star$; since the incidence angle equals the reflection angle (Snell's law), the reflected ray makes the angle 2α with (Oy). The slope of the reflected ray is thus $-1/\tan(2\alpha) = (\tan^2 \alpha - 1)/(2 \tan \alpha)$. The equation of the reflected ray thus reads:

$$y - y^\star = \left(\frac{4a^2 x^{\star 2} - 1}{4ax^\star} \right)(x - x^\star).$$

It cuts the y-axis in $(0, 1/4a)$, regardless of the value of $x^\star$. We thus have $f = 1/4a$ for the focal length of the mirror. Figure 2.7(b) illustrates this property of the parabolic mirror: all rays coming from infinity (along the axis) converge at the focus F. The parabolic mirror is the only one to be rigorously stigmatic[3]. The reader is invited to show this property, by performing the above calculations for an arbitrary mirror shape, given by a curve $y = f(x)$. The stigmatism condition implies that the position of the focus does not depend on the point $x^\star$ where the ray hits the mirror. Mathematically, this translates into an ordinary differential equation for $f(x)$. One finds that only a parabola fulfills the stigmatism condition.

---

[3] For rays coming from infinity and parallel to the mirror axis.

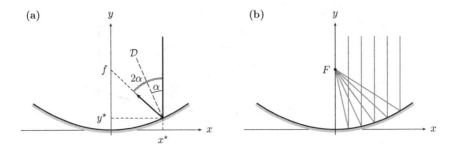

Fig. 2.7   (a) *Determination of the focus F of an optical parabolic mirror.* (b) *The rays coming from infinity and parallel to the axis Oy of the mirror all converge to F.*

(1) We apply Newton's second law with the initial conditions given in the text. The abscissa x_{n+1} of the impact point on the mirror fulfills:

$$ax_{n+1}^2 = -\frac{1}{2}gt^{\star 2} + v_{y,n}t^\star + ax_n^2 \qquad (2.9)$$

$$x_{n+1} = x_n + v_{x,n}t^\star \qquad (2.10)$$

where $t^\star$ is the time at which the collision takes place. By combining Eqs (2.9) and (2.10), we find

$$a(x_{n+1} + x_n)v_{x,n} = (v_{y,n} - gt^\star/2), \qquad (2.11)$$

which gives, by eliminating $t^\star$ thanks to Eq. (2.10), the result:

$$x_{n+1} = \frac{1}{g + 2av_{x,n}^2}\left[2v_{x,n}v_{y,n} + x_n\left(g - 2av_{x,n}^2\right)\right].$$

(2) The components of v_- are ($v_x = v_{x,n}, v_y \equiv v_{y,n} - gt^\star$). Let i be the unit vector tangent to the mirror at the impact point x_{n+1}, and j a unit vector that is orthogonal to i (see Fig. 2.8). One has

$$i = \frac{1}{\sqrt{1 + 4a^2 x_{n+1}^2}}\begin{pmatrix} 1 \\ 2ax_{n+1} \end{pmatrix}$$

and

$$j = \frac{1}{\sqrt{1 + 4a^2 x_{n+1}^2}}\begin{pmatrix} -2ax_{n+1} \\ 1 \end{pmatrix}.$$

We rewrite v_- in this basis:

$$v_- = (v_- \cdot i)i + (v_- \cdot j)j.$$

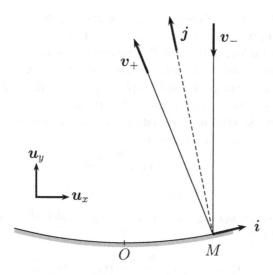

Fig. 2.8 *Elastic bounce of the particle on the parabolic mirror. The collision occurs at point M, with coordinates (x_{n+1}, ax_{n+1}^2) in the coordinate system $(O, \boldsymbol{u}_x, \boldsymbol{u}_y)$.*

The bounce changes the sign of the component along $\boldsymbol{j}$ in such a way that the velocity of the atom after the collision $\boldsymbol{v}_+$ simply reads:

$$\boldsymbol{v}_+ = (\boldsymbol{v}_- \cdot \boldsymbol{i})\boldsymbol{i} - (\boldsymbol{v}_- \cdot \boldsymbol{j})\boldsymbol{j}.$$

We deduce the expression of $v_{x,n+1} = \boldsymbol{v}_+ \cdot \boldsymbol{u}_x$:

$$v_{x,n+1} = \frac{1}{1 + 4a^2 x_{n+1}^2} \left[v_{x,n} \left(1 - 4a^2 x_{n+1}^2 \right) + 4ax_{n+1}v_y \right].$$

(3) From $v_{x,n} = 0$ we get $x_{n+1} = x_n$. Using the above discussion, we thus have:

$$v_{x,n+1} = \frac{4ax_n v_y}{1 + 4a^2 x_{n+1}^2}.$$

Equation (2.11) allows us to deduce that $v_{y,n} = gt^\star/2$, and thus $v_y = v_{y,n} - gt^\star = -v_{y,n} = -gt^\star/2$. After the $(n+1)^{\text{th}}$ bounce, the trajectory of the atom has the equation $x = x_n + v_{x,n+1}t$ and the atom cuts the ordinate axis after a time $\tilde{t} = -x_n/v_{x,n+1}$, at the position

$$y = -\frac{1}{2}g\tilde{t}^2 + ax_n^2 + v_{y,n}\left(-\frac{x_n}{v_{x,n+1}}\right),$$

or

$$y = \frac{1}{4a} + 2ax_n^2 - \frac{g}{32a^2 v_{y,n}^2}(1 + 4a^2 x_{n+1}^2)^2.$$

The focal length, obtained for $v_{y,n} \to \infty$ and $ax_n \ll 1$, is thus $f = 1/4a$, just as in optics. The assumption $ax_n \ll 1$ amounts to requiring that the atoms explore only a small transverse region as compared to the radius of curvature of the mirror. The approximation $v_{y,n} \to \infty$ is the analogue of the one performed in optics in order to determine the focus: one studies a "ray" which is parallel to the axis.

(4) Within the paraxial approximation ($ax_n \ll 1$ and $av_{x,n}^2 \ll g$), the above results can be rewritten as:

$$x_{n+1} = x_n + \frac{2v_{x,n}v_{y,n}}{g},$$

$$v_{x,n+1} = v_{x,n} + 4ax_{n+1}v_y. \tag{2.12}$$

But $v_y = v_{y,n} - g(x_{n+1} - x_n)/v_{x,n} \simeq -v_{y,n}$, which allows us to rewrite Eq. (2.12) as:

$$v_{x,n+1} = \left(1 - \frac{8av_{y,n}^2}{g}\right)v_{x,n} - 4av_{y,n}x_n.$$

In the paraxial regime, v_y is not affected by the bounces, and one replaces, in what follows, $v_{y,n}$ by v_y:

$$\begin{pmatrix} x_{n+1} \\ v_{x,n+1} \end{pmatrix} = \begin{pmatrix} 1 & 2v_y/g \\ -4av_y & 1 - 8av_y^2/g \end{pmatrix} \begin{pmatrix} x_n \\ v_{x,n} \end{pmatrix} = \mathsf{M} \begin{pmatrix} x_n \\ v_{x,n} \end{pmatrix}.$$

We have $\det \mathsf{M} = 1$, and therefore the trajectory is stable transversally if the eigenvalues are complex conjugates one of the other, with unity modulus[4], i.e. if $|\mathrm{Tr}(\mathsf{M})| < 2$. If we introduce the height of a bounce $H = v_y^2/2g$, the stability condition reads simply $H < f$.

The matrix formulation that we have used here is inspired by the matrix formulation of geometrical optics within the Gauss approximation. Such a mirror for atoms was realized for the first time in 1993 with cesium atoms[5]. In this pioneering study, the atoms could be observed to bounce up to 13 times in the cavity. In such an experiment, the mirror is realized using an evanescent wave at the surface of a glass prism. This idea had been proposed theoretically in 1982 by R.J. Cook and R.K. Hill[6]. This system

---

[4]In the opposite case, as the product of the eigenvalues is 1, there would necessarily be an eigenvalue with a modulus higher than one, and, when iterating the action of M, we would obtain an exponential divergence of $(x_n, v_{x,n})$.

[5]C.G. Aminoff, A.M. Steane, P. Bouyer, P. Desbiolles, J. Dalibard, and C. Cohen-Tannoudji, *Cesium atoms bouncing in a stable gravitational cavity*, Phys. Rev. Lett. **71**, 3083 (1993).

[6]R.J. Cook and R.K. Hill, *An electromagnetic mirror for neutral atoms*, Opt. Commun. **43**, 258 (1982).

allows for the realization of a cavity in which one of the "walls" is made by light, and the other is provided by gravity. This is the reason why one speaks of *gravitational cavities*. The studies in this direction had primarily metrologic motivations: in such a setting, the atoms remain most of the time unperturbed by light. Nowadays, more advanced techniques, such as *magic optical lattices*[7], are actively explored in this field to achieve the same goal, but with much longer storage times.

[7]H. Katori, M. Takamoto, V.G. Palchikov, and V.D. Ovsiannikov, *Ultrastable optical clock with neutral atoms in an engineered light shift trap*, Phys. Rev. Lett. **91**, 173005 (2003), also freely available on the internet at the address http://arxiv.org/abs/physics/0309043.

Chapter 3

Friction

3.1 Energy balance for forced oscillations *

We consider a physical system described by the following differential equation:

$$\ddot{x} + 2\lambda\dot{x} + \omega_0^2 x = \frac{f}{m}\cos(\gamma t), \qquad (3.1)$$

with $\lambda > 0$.

(1) Give at least one example of such a physical system.
(2) Show that the general solution of Eq. (3.1) can be put in the form:

$$x \simeq ae^{-\lambda t}\cos(\omega_0 t + \varphi) + b\cos(\gamma t + \delta), \qquad (3.2)$$

if $\lambda \ll \omega_0$. Interpret the two terms of this solution, and give the explicit expression of b and $\tan\delta$.
(3) We consider forced oscillations. What happens to Eq. (3.2)?
(4) Show that under forced oscillations, one has

$$\left\langle 2\lambda\dot{x}^2 \right\rangle = \left\langle \dot{x}\frac{f}{m}\cos(\gamma t) \right\rangle, \qquad (3.3)$$

where the average $\langle \cdots \rangle$ is taken over a period $T = 2\pi/\gamma$ of the forced oscillations. Give the energetic meaning of this relation.

Solution

(1) Let us give two classical examples of physical situations in which Eq. (3.1) appears.

- In mechanics, Eq. (3.1) describes the motion of a mass m attached to a spring with a spring constant k, that undergoes forced oscillations under the action of the force $f \cos(\gamma t)$, in the presence of a fluid friction force (term proportional to $\dot{x}$). The variable x denotes the position of the mass. The angular frequency ω_0 is related to the spring constant by $\omega_0^2 = k/m$.

- In electricity, Eq. (3.1) describes the time evolution of the electric charge in a RLC circuit that consists of a resistor (R), an inductor (L), and a capacitor (C) connected in series, with a sinusoidal time-dependent voltage as a power source. The expression for the angular frequency is here given by $\omega_0 = 1/\sqrt{LC}$, and the damping term by $\lambda = R/(2L)$. Such an electrical circuit is equivalent to the mechanical system where the resistor plays the role of the fluid friction term, the inductor acts as an inertial term, and the capacitor as a restoring force.

More generally, there exist many situations in physics where the same mathematical equations with different meanings for the variables describe phenomena that originate from very different fields. For instance, some acoustics problems can be recast in a mathematical form identical to one encountered in general relativity!

(2) The general solution $x(t) = x_{\text{ho}}(t) + x_{\text{part}}(t)$ of Eq. (3.1) is the sum of a transient solution $x_{\text{ho}}(t)$ (the solution for the free damped harmonic oscillator) that depends on initial conditions, and of a steady-state solution $x_{\text{part}}(t)$, that is a particular solution of the non-homogeneous differential equation and that is independent of initial conditions.

The transient solution is readily obtained by plugging a solution of the form $x_{\text{ho}} \propto e^{rt}$ into Eq. (3.1) with $f = 0$. The parameter r is solution of the following quadratic equation:

$$r^2 + 2\lambda r + \omega_0^2 = 0.$$

The discriminant Δ of this equation reads $\Delta = 4(\lambda^2 - \omega_0^2) \simeq -4\omega_0^2$, since, according to the text, we restrict our analysis to the case $\lambda \ll \omega_0$. To the lowest order in λ/ω_0, one has $r_\pm = -\lambda \pm i\omega_0$, and the damping is responsible for the exponential decay of the amplitude of the oscillations. If we continue the expansion up to the second order in λ/ω_0, we find that the damping term modifies the frequency of oscillation. In the following, we neglect this effect and restrict to the first-order corrections. As the solution is the one of a second-order linear differential

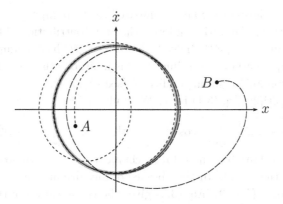

Fig. 3.1 *Trajectory in phase space* $(x, \dot{x})$ *of two particles, located initially at A (short dashed line) and at B (long dashed line). The driven solution corresponds to a circular trajectory (gray circle). The trajectories of the two particles collapse onto the circle after a sufficiently large time (large with respect to the damping time* $1/\lambda$ *associated with the fluid friction force).*

equation, there are two constants of integration: the amplitude a, and the phase φ.

To find a particular solution, we use the complex variables method and search for a solution of the equation

$$\ddot{X} + 2\lambda \dot{X} + \omega_0^2 X = \frac{f}{m} e^{i\gamma t}$$

of the type $X = Be^{i\gamma t}$. We find:

$$B = \frac{f}{m} \frac{1}{\omega_0^2 - \gamma^2 + 2i\lambda\gamma}.$$

The amplitude B is then written in the modulus-phase form as $B = be^{i\delta}$. One readily obtains the equation satisfied by the modulus and phase:

$$b = \frac{f}{m} \frac{1}{\sqrt{(\omega_0^2 - \gamma^2)^2 + 4\lambda^2\gamma^2}} \quad \text{and} \quad \tan\delta = \frac{2\lambda\gamma}{\gamma^2 - \omega_0^2}.$$

The particular solution we were looking for is given by the real part of X: $x_{\text{part}} = b\cos(\gamma t + \delta)$. We conclude that the most general form for the solution of Eq. (3.1) is:

$$x \simeq ae^{-\lambda t}\cos(\omega_0 t + \varphi) + b\cos(\gamma t + \delta). \tag{3.4}$$

(3) After an evolution time large compared to the damping time $1/\lambda$ associated with the fluid friction force, the only contribution of the solution that remains is $x_{\text{part}}(t)$. Indeed, the term $x_{\text{ho}}(t)$ has an amplitude that decays exponentially with time. In this limit, one therefore loses all the information about initial conditions (see Fig. 3.1).

(4) Let us multiply Eq. (3.1) by $\dot{x}$. We find:

$$\frac{\mathrm{d}E}{\mathrm{d}t} + 2\lambda\dot{x}^2 = \frac{f}{m}\dot{x}\cos\gamma t \qquad \text{with} \qquad E = \frac{1}{2}\dot{x}^2 + \frac{1}{2}\omega_0^2, \qquad (3.5)$$

which is nothing but the total mechanical energy of the oscillator (divided by the mass). In the absence of any driving force ($f = 0$) and friction force ($\lambda = 0$), Eq. (3.5) gives the conservation of the total energy that contains two terms, the potential and the kinetic terms. In the absence of the driving force ($f = 0$), Eq. (3.5) describes the decrease of the total mechanical energy due to the friction force. The driving force acts as a source term in Eq. (3.5), while the friction term accounts for dissipation. In the following, we show that, on average and for a sufficiently large time ($t \gg 1/\lambda$, i.e. in the steady-state regime), the gain in energy from the source is exactly dissipated by the friction force. In order to prove Eq. (3.3) we thus need to show that $\langle \mathrm{d}E/\mathrm{d}t \rangle = 0$. Let us calculate $\mathrm{d}E/\mathrm{d}t$ in the steady-state regime:

$$\frac{\mathrm{d}E}{\mathrm{d}t} = \dot{x}\ddot{x} + \omega_0^2\dot{x}x = \frac{\gamma b^2}{2}(\gamma^2 - \omega_0^2)\sin\left[2(\gamma t + \delta)\right].$$

Now,

$$\int_0^{2\pi/\gamma} \sin\left[2(\gamma t + \delta)\right]\,\mathrm{d}t = 0,$$

which gives the expected result. We conclude that in the steady-state regime, all the energy given by the source is dissipated on average (over one period) by the friction force.

3.2 Solid friction versus fluid friction *

(1) Recall the empirical laws of solid friction. Introduce the coefficients of *static* and *kinetic* friction.

(2) We consider a brick of mass m lying on a table and attached to a fixed point by a spring scale (see Fig. 3.2). The spring constant is denoted by k. We assume that the spring is elongated enough to set the brick into motion and we denote by $x_0 > 0$ the extension of the

spring scale with respect to its equilibrium position (at rest, without any constraint). The brick starts at $t = 0$ with a vanishing initial velocity. The solid friction is characterized by a friction coefficient f (we do not distinguish here between the static and kinetic coefficients). Propose a graphical method to solve the motion of the brick in phase space (i.e. a plane with the position $X = x$ as the abscissa axis, and the velocity $Y = (dx/dt)/\omega$, where ω is the angular frequency of the mass-spring system, as the ordinate axis).

(3) Compare the phase-space trajectory with that of an oscillator damped by a fluid friction force.

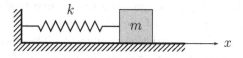

Fig. 3.2 *Brick attached to a spring and subjected to solid friction.*

Solution

(1) When two solid bodies, such as the brick and the table, interact, there are contact forces between them. The empirical laws of solid (or *dry*) friction give information on the components of the reaction force exerted by the substrate onto the solid. The contact force exerted by the table onto the brick has a component perpendicular to the surface, the normal force N, and a component parallel to the surface, the friction force T. In the absence of sliding, there exists an inequality between those two components: $|T| \leqslant f_s|N|$, where f_s is a dimensionless parameter referred to as the *coefficient of static friction*. In the presence of sliding, the norm of the tangential component is fixed and proportional to the normal component: $|T| = f_k|N|$. The coefficient f_k is the *coefficient of kinetic friction*. The direction of the tangential component is opposite to the velocity, since, by definition, a friction force acts against the motion. We recall that the friction coefficients necessarily fulfill the inequality $f_k \leqslant f_s$. Indeed, let us assume that this inequality is not fulfilled. In this case, when the force applied to set the solid into motion is sufficient to overcome static friction, the friction would suddenly increase and the solid would remain at rest until the force

reaches the value of the kinetic friction force. Such a situation yields a clear contradiction.

(2) To draw the phase portrait, one needs to study typical trajectories. The direction of the tangential force depends on sign of the velocity. We therefore have to solve the equation of motion for two separate cases: $\dot{x} < 0$ and $\dot{x} > 0$.

Let us start with $\dot{x} < 0$. According to Newton's law, the equation of motion along the x axis reads:

$$m\frac{\mathrm{d}^2 x}{\mathrm{d}t^2} = -kx + T.$$

This equation contains two terms: the first one accounts for the force exerted by the spring scale, and the second one tends to slow down the brick under the action of the friction force. If $\dot{x} < 0$, $T > 0$ and $T = fN$, where f is the friction coefficient. The normal component of the contact force compensates for the weight so that $N = mg$. We deduce the equation of motion of the brick in the case $\dot{x} < 0$:

$$m\frac{\mathrm{d}^2 x}{\mathrm{d}t^2} = -kx + fmg.$$

This equation can be rewritten in the form:

$$\frac{\mathrm{d}^2 x}{\mathrm{d}t^2} + \omega^2(x - x_e) = 0, \tag{3.6}$$

where we have introduced the angular frequency $\omega^2 = k/m$ associated with the harmonic restoring force exerted by the spring scale, and a characteristic length $x_e = fmg/k$. The initial extension of the spring scale fixes the initial condition $x(0) = x_0$. If $x_0 < x_e$, the brick remains at rest since the friction force is larger than that exerted by the spring scale. To set the brick in motion, the inequality $x_0 > x_e$ has to be fulfilled. To draw a trajectory on the phase portrait, one needs to know the velocity $\dot{x} = \mathrm{d}x/\mathrm{d}t$ as a function of the position x. This relation can be obtained from Eq. (3.6) by multiplying it by $\dot{x}$:

$$\frac{\mathrm{d}x}{\mathrm{d}t}\frac{\mathrm{d}^2 x}{\mathrm{d}t^2} + \omega^2(x - x_e)\frac{\mathrm{d}x}{\mathrm{d}t} = 0.$$

The integration over time of the latter equation gives:

$$\frac{m}{2}\left(\frac{\mathrm{d}x}{\mathrm{d}t}\right)^2 + \frac{1}{2}m\omega^2(x - x_e)^2 = \text{const.} \tag{3.7}$$

Note that this conserved quantity differs from the total mechanical energy $m\dot{x}^2/2 + m\omega^2 x^2/2$; the latter is not conserved since the friction force induces dissipation. Dividing Eq. (3.7) by $m\omega^2/2$, we can rewrite the equation of motion in terms of the variables (X, Y) in the form:

$$Y^2 + (X - x_e)^2 = R_1^2, \tag{3.8}$$

where R_1 is a constant. Equation (3.8) is the equation of a circle whose center is at $X = x_e$ and whose radius is R_1. However, one should keep in mind that this result is valid only under the assumption $\dot{x} < 0$. In the plane (X, Y), the trajectory begins by a half-circle centered at x_e in the region $Y < 0$. The value of the radius R_1 is dictated by the initial conditions $x = x_0$ and $\dot{x}(0) = 0$: we have $R_1 = x_0 - x_e > 0$. For $\dot{x} > 0$, the same reasoning yields:

$$Y^2 + (X + x_e)^2 = R_2^2. \tag{3.9}$$

At the point of coordinates $X_1 = x_e - R_1 = 2x_e - x_0$ and $Y = 0$, Eqs (3.8) and (3.9) are both fulfilled. If $-x_e \leqslant X_1 \leqslant x_e$, the restoring force cannot counteract the friction force and the brick remains at rest. Otherwise, $X_1 \leqslant -x_e$, and we have $R_2 = x_0 - 3x_e > 0$. In the plane (X, Y), the trajectory is a half-circle centered on $-x_e$. At each change of sign of the velocity, one has to check if the brick position is within the interval $[-x_e, x_e]$. If this is the case, the brick stops at the position where its velocity vanishes. If not, the trajectory continues by another half-circle centered at the other extremity of the interval $[-x_e, x_e]$. By iteration, we obtain a graphical solution as a succession of half-circles [see Fig. 3.3(a)].

(3) The viscous drag force is the force exerted by a fluid on a moving object immersed in it. This force is proportional to the velocity of the object and acts in a direction opposite to the velocity: $F = -m\lambda\dot{x}$. The equation of motion of the object is given by Newton's second law:

$$m\frac{d^2x}{dt^2} = -kx - m\lambda\frac{dx}{dt}. \tag{3.10}$$

Let us perform as previously the calculation of the expression of the velocity as a function of the position. We multiply Eq. (3.10) by dx/dt. In terms of the variables X and Y, we obtain:

$$\frac{d}{dt}\left(X^2 + Y^2\right) = -2\lambda Y^2 < 0. \tag{3.11}$$

We find that the total mechanical energy decreases through the action of the viscous drag force. It is clear from Eq. (3.11) that the radius of

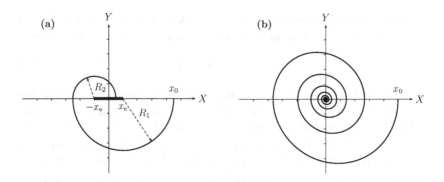

Fig. 3.3 (a) *Phase portrait of an oscillator damped by a solid friction force. The trajectory is a succession of half-circles centered at $\pm x_e$ ($x_0 = 0.9$, $x_e = 0.2$). (b) Phase portrait of an oscillator damped by a viscous drag force. The trajectory is a spiral with a decreasing radius ($x_0 = 1$, $\omega = 1$, $\lambda = 0.2$).*

the trajectory continuously decreases as a function of time. To find the exact form of the trajectory we need to solve Newton's equation:

$$\frac{d^2x}{dt^2} + \lambda\frac{dx}{dt} + \omega^2 x = 0,$$

with $\lambda > 0$. This second-order linear differential equation can be solved easily by searching solutions of the form $x \sim e^{rt}$. We find that the exponent r is the root of the quadratic equation $r^2 + \lambda r + \omega^2 = 0$. The discriminant is equal to $\Delta = \lambda^2 - 4\omega_0^2$. Let us assume, for the sake of simplicity, that the viscous drag force has a relatively small effect on the motion of the object. This happens when the condition $\lambda \ll \omega_0$ is fulfilled. In this case, the discriminant is negative, and the two complex solutions for the exponent r are:

$$r_\pm = \frac{1}{2}\left(-\lambda \pm i\sqrt{4\omega^2 - \lambda^2}\right).$$

The general form of the solution of Eq. (3.10) is:

$$x(t) = e^{-\lambda t/2}\left[A\cos\left(\sqrt{4\omega^2 - \lambda^2}t/2\right) + B\sin\left(\sqrt{4\omega^2 - \lambda^2}t/2\right)\right],$$

where A and B are constants determined by the initial conditions, that are $x(0) = x_0$ and $\dot{x}(0) = 0$. We finally find

$$A = x_0 \quad \text{and} \quad B = \frac{\lambda x_0}{2\sqrt{4\omega^2 - \lambda^2}}.$$

We have plotted on Fig. 3.3(b) an example of a trajectory that spirals towards $x = 0$. We emphasize the difference from the dry friction force addressed in the first part of the exercise. The position for which the brick stops when it experiences a viscous drag force is $x = 0$, while with the solid friction force we can only say that the final position is within the interval $[-x_e, x_e]$, and its exact value depends on the initial conditions.

♦ **Remark.** Taking into account the fact that the coefficient of kinetic friction is lower than the coefficient of static friction enriches the dynamics considerably. For instance, if one pulls, by means of a spring scale, a solid which is subjected to solid friction, one observes the so-called *stick-slip* motion, made of two phases that periodically alternate: one observes first that the solid is at rest (it *sticks* to the substrate, so to say) while the spring elongates, and then the solid moves (*slips*) suddenly (since the static friction does not compensate any more for the force exerted by the spring), until it comes to rest again. Such a periodic motion, with successive phases where stress accumulates and is then released, can be observed in phenomena ranging from the squeaking of a piece of chalk on a blackboard, to the excitation of the strings of a violin by the bow, and the triggering of earthquakes.

3.3   Fluid friction and delayed response **

In this exercise, we consider a particle whose velocity $v = dx/dt$ evolves according to the equation:

$$\frac{dv}{dt} = -\gamma v + F(x), \qquad (3.12)$$

with $\gamma > 0$.

(1) Give the physical meaning of this equation.
(2) We assume that $F(x) = F$ does not depend on the position x. Calculate the steady-state velocity.
(3) In this question, we consider the general case for which F depends on x. Furthermore, we assume that the velocity evolves, under the action of F, on a timescale that is long in comparison with the friction timescale γ^{-1}. Show that, in the quasi-steady-state regime, the velocity $v(t)$ has the same mathematical form as in the previous question, but with the force F taken at the position $x(\tilde{t})$ at a time $\tilde{t}$ delayed with respect to the current time t. Give the expression of $\tilde{t}$ as a function of t and γ.

Solution

(1) Equation (3.12) is the equation of motion of a particle that experiences

a fluid friction force and a position dependent force (per unit mass) $F(x)$. The viscous damping time is given by γ^{-1}.

(2) In this case, the steady-state regime is readily obtained by setting $dv/dt = 0$. We find as a result of the competition between F and the drag $-\gamma v$ an asymptotic terminal velocity which is constant: $v_{\text{lim}} = F/\gamma$. A ball falling in a vertical tube that contains a viscous liquid therefore reaches a limiting velocity dictated by the viscous drag. Actually, this is true only for objects moving through the fluid at sufficiently low speeds so that turbulence is negligible. For this specific example $F = g$, where g denotes the acceleration due to gravity.

(3) If F depends on the position x, the answer is more subtle. We can always integrate Eq. (3.12) formally:

$$v(t) = v(0)\,\mathrm{e}^{-\gamma t} + \int_0^t \mathrm{e}^{-\gamma(t-t')} F[x(t')]\,\mathrm{d}t'. \qquad (3.13)$$

For a time t' sufficiently large as compared to the viscous damping time γ^{-1}, but small with respect to the timescale of evolution of the velocity under the action of the force F, one can write $x(t') \simeq x(t) + (t'-t)v$, where v is the velocity at time t. The force can therefore be expanded at first order:

$$F[x(t')] \simeq F[x(t) + (t'-t)v] \simeq F[x(t)] + (t'-t)v\frac{\partial}{\partial x}F[x(t)].$$

By inserting this expression for the force into Eq. (3.13), we find after integration and neglecting all terms proportional to $\mathrm{e}^{-\gamma t}$ (because they are damped out in the quasi-steady-state regime):

$$\gamma v(t) = F[x(t)] - v\gamma^{-1}\frac{\partial}{\partial x}F[x(t)].$$

This expression can be considered as resulting from a first-order expansion with respect to the position x:

$$\gamma v(t) \simeq F[x(t) - v\gamma^{-1}] \simeq F[x(t-\gamma^{-1})] = F[x(\tilde{t})]$$

with $\tilde{t} = t - \gamma^{-1}$. This demonstrates the result suggested in the text: the response at time t depends on the position at an earlier time $\tilde{t}$ because of the delay introduced by the drag force. This link between friction and delayed response plays an important role in the interpretation of physical phenomena where fluid friction appears. This is, for instance, the case of laser cooling of atoms[1].

---

[1] An example of such a laser cooling mechanism is studied in the problem entitled "Doppler cooling" (Problem 9.3).

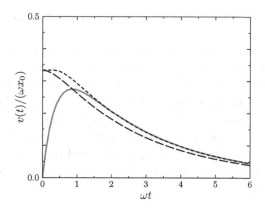

Fig. 3.4 *Velocity of an overdamped harmonic oscillator ($\gamma = 3\omega$): exact calculation (solid gray line), static steady state approximation (long-dashed line) and the delayed response approach (short-dashed line).*

In order to illustrate the quality of the approximation obtained by the delayed response, let us investigate a specific example for which the analytical solution is known. We consider a position-dependent restoring force of the form $F(x) = -\omega^2 x$, assuming $\gamma \gg \omega$. One therefore needs to solve the equation of motion of an oscillator in the overdamped regime. The velocity $v(t)$ of a particle initially at rest and at the position x_0 is plotted in Fig. 3.4 for $\gamma = 3\omega$ using three different approaches:

- exact analytical solution (solid gray line);
- using the static steady-state approximation $v(t) \simeq -F\left[x(t)\right]/\gamma$ (long-dashed line);
- using the delayed response calculation $v(t) \simeq -F\left[x(\tilde{t})\right]/\gamma$ with $\tilde{t} = t - 1/\gamma$ (short-dashed line).

We clearly observe that the delayed response approach gives an excellent account of the motion in its validity domain $t \gg 1/\gamma$.

We emphasize that the approach of the delayed response developed in this exercise can be carried out on all equations of the same type as Eq. (3.12)[2]. This means that such a solution occurs in a large variety of problems, such as in electronics and signal processing for example.

---

[2]And even, in a simpler way, in the case where the "force" F is a given time-dependent function (i.e., it does not depend on x explicitly): this is in particular a situation one has to face in electronics, with first-order filters for instance.

3.4   Absence of dissipation in a system with friction **

In this exercise, one shows a counter-intuitive result: coupling a damped oscillator to another, undamped oscillator can lead in some cases to the absence of dissipation in the system made of both oscillators.

Consider the system shown in Fig. 3.5: two points, (1) and (2), of mass m and positions x_i ($i \in \{1, 2\}$), are both connected to fixed points by identical springs (spring constant k) and are connected together by a spring of stiffness K. Point (1) is also submitted to a drag force $-m\gamma\dot{x}_1$. One studies the system when particle (1) is driven by a periodic force $F(t) = F_0 \cos \omega t$.

(1) Write down the equations of motion for $x_1(t)$ and $x_2(t)$. Hint: introduce the quantities $\omega_0^2 = (k + K)/m$ and $\Omega^2 = K/m$.
(2) Calculate, in the steady-state regime, the average power $\langle P_1 \rangle$ supplied by the operator acting on particle (1) (the time average being calculated over one period of the motion).
(3) Assume that there is no coupling between the two particles (i.e. $\Omega = 0$). Plot $\langle P_1 \rangle$ versus $\delta = \omega - \omega_0$ in the regime $\delta \ll \omega_0$.
(4) Show that the presence of the coupling Ω, whatever its strength, makes $\langle P_1 \rangle$ vanish at resonance. Interpret. Give the expression of $\langle P_1 \rangle(\delta)$ in the case $\delta \ll \omega_0$. Show that the frequency range over which dissipation is significantly reduced is on the order of Ω^2/ω_0. Plot $\langle P_1 \rangle(\delta)$ for fixed γ, for various values of the coupling Ω.

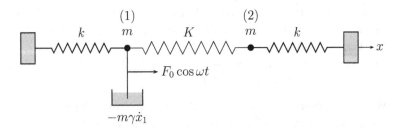

Fig. 3.5   *System of two coupled oscillators, of which one is damped.*

Solution

(1) By applying Newton's second law to mass (1) and to mass (2), respectively, one gets:

$$\begin{cases} m\ddot{x}_1 = F_0 \cos\omega t - kx_1 - m\gamma\dot{x}_1 - K(x_1 - x_2), \\ m\ddot{x}_2 = -kx_2 - K(x_2 - x_1). \end{cases}$$

Using the notations $\omega_0^2 = (k + K)/m$ and $\Omega^2 = K/M$, we can rewrite the above equations in the form:

$$\begin{cases} \ddot{x}_1 + \gamma\dot{x}_1 + \omega_0^2 x_1 = \dfrac{F_0}{m} \cos\omega t + \Omega^2 x_2, \\ \\ \ddot{x}_2 + \omega_0^2 x_2 = \Omega^2 x_1. \end{cases}$$

(2) One is interested in the steady-state regime at (angular frequency) ω, so one uses the complex notation:

$$F_0 \cos\omega t \longrightarrow F_0 e^{i\omega t},$$

$$x_1(t) \longrightarrow \tilde{x}_1 e^{i\omega t},$$

$$x_2(t) \longrightarrow \tilde{x}_2 e^{i\omega t}.$$

Upon substitution in the previous set of equations, one gets the following set of linear equations:

$$\begin{cases} (\omega_0^2 - \omega^2 + i\omega\gamma)\tilde{x}_1 - \Omega^2\tilde{x}_2 = F_0/m, \\ -\Omega^2\tilde{x}_1 + (\omega_0^2 - \omega^2)\tilde{x}_2 = 0. \end{cases}$$

The expression for $\tilde{x}_1$ thus reads:

$$\tilde{x}_1 = \frac{F_0}{m} \frac{\omega_0^2 - \omega^2}{(\omega_0^2 - \omega^2 + i\omega\gamma)(\omega_0^2 - \omega^2) - \Omega^4}.$$

The power supplied by the operator reads $P_1(t) = \boldsymbol{F}(t) \cdot \boldsymbol{v}(t) = F_0(t)\dot{x}_1(t)$. To obtain the average power over one period, one just needs to calculate

$$\langle P_1 \rangle = \mathrm{Re}\left(\frac{1}{2}\tilde{x}_1 F_0^\star\right),$$

where $z^\star$ denotes the complex conjugate of z, and $\mathrm{Re}(z)$ the real part of z. We find:

$$\langle P_1 \rangle = \frac{F_0^2}{2m} \frac{\omega^2\gamma\left(\omega_0^2 - \omega^2\right)^2}{\left[(\omega_0^2 - \omega^2)^2 - \Omega^4\right]^2 + \omega^2\gamma^2\left(\omega_0^2 - \omega^2\right)^2}. \tag{3.14}$$

(3) For a vanishing coupling ($\Omega = 0$), the previous expression becomes

$$\langle P_1 \rangle = \frac{F_0^2}{2m} \frac{\omega^2 \gamma}{\left(\omega_0^2 - \omega^2\right)^2 + \omega^2 \gamma^2}.$$

If the detuning δ is small compared to ω_0, one can write $\omega_0^2 - \omega^2 = (\omega_0 - \omega)(\omega_0 + \omega) \simeq 2\omega\delta$, and the expression of the dissipated power becomes:

$$\langle P_1 \rangle \simeq \frac{F_0^2}{2m\gamma} \frac{1}{1 + 4\delta^2/\gamma^2}.$$

This is a Lorentzian resonance curve, with a full width at half-maximum γ [see Fig. 3.6 (a)].

(4) In the presence of a coupling $\Omega \neq 0$, whatever its strength, Eq. (3.14) immediately shows that one has $\langle P_1 \rangle(\omega_0) = 0$. The dissipated power vanishes, because in the stationary regime particle (1) stands still. It is important to understand that during the transient regime, particle (1) does oscillate under the action of the force $F_0 \cos \omega t$, and *via* the coupling Ω, induces a motion in particle (2). But on resonance, when the stationary regime is reached, mass (2) is the only one moving, with a phase such that the force $K(x_2 - x_1)\boldsymbol{u}_x$ it exerts onto (1) exactly compensates for the force $F_0 \cos \omega t \, \boldsymbol{u}_x$.

Expression (3.14) considerably simplifies in the case $\delta \ll \omega_0$, since one has, replacing $\omega^2 - \omega_0^2$ by the approximate value $2\delta\omega_0$:

$$\langle P_1 \rangle \simeq P_0 \frac{1}{1 + \left(\dfrac{\Omega^4 - 4\omega_0^2\delta^2}{2\omega_0^2\delta\gamma}\right)^2},$$

where $P_0 = F_0^2/(2m\gamma)$. In order to study the variations of this function when δ varies, one can calculate its derivative and study its sign, but this is not necessary. Indeed, the function $X \to 1/(1 + X^2)$ is strictly decreasing on the interval $[0, \infty[$. Its maximum (with a value of 1) is thus reached for $X = 0$, and its minimum (with a value of zero) for $X \to \infty$. But, as a function of δ, $X \equiv (\Omega^4 - 4\omega_0^2\delta^2)/(2\omega_0^2\delta\gamma)$ vanishes only for $\delta = \pm\Omega^2/(2\omega_0)$ (there, $\langle P_1 \rangle$ is maximal and has the value P_0), and becomes infinite only when $\delta \in \{0, \pm\infty\}$. The even function $\langle P_1 \rangle(\delta)$ therefore increases from 0 to P_0 when δ varies from 0 to $\Omega^2/(2\omega_0)$, and then decreases towards a vanishing value at $\delta = \infty$. A plot of $\langle P_1 \rangle(\delta)$ is shown on Fig. 3.6. For a weak coupling (i.e. when $\Omega^2 \ll \gamma\omega_0$), the effect of the coupling is to introduce in the center of the resonance curve a very narrow peak in which dissipation is strongly

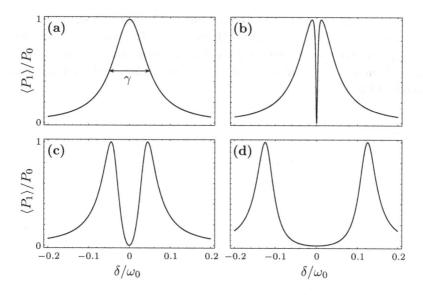

Fig. 3.6 $\langle P_1 \rangle (\delta)$, *for* $\gamma = \omega_0/10$ *and various values of the coupling* Ω: (a) $\Omega = 0$, (b) $\Omega = 0.1\,\omega_0$, (c) $\Omega = 0.3\,\omega_0$, (d) $\Omega = 0.5\,\omega_0$.

reduced [Fig. 3.6 (b)]. For a strong coupling ($\Omega^2 \gg \gamma\omega_0$) one has essentially two peaks centered at $\omega_0 \pm \Omega^2/(2\omega_0)$, with a width $\sim \gamma/2$ [Fig. 3.6 (d)].

▶ **Further reading.** The system studied in this exercise is a simple mechanical analogue of a phenomenon that has attracted a lot of interest in recent years in atomic physics, and that is called *Electromagnetically Induced Transparency* (EIT): under certain conditions, a medium which usually is absorbing light of a given frequency (corresponding to a transition between two atomic levels) can become transparent for this light beam (called the *probe*), when one adds another beam (the *pump*) which induces a coupling to a third atomic level. One of the spectacular applications of this phenomenon consists in decreasing considerably (or even zeroing) the velocity of a light pulse propagating in the medium. One can indeed show that for a very narrow peak of transparency, such as that of Fig. 3.6 (b) for instance, one has a very rapid variation of the refractive index of the medium with the wavelength λ of the light; this implies that the group velocity of a wavepacket of light can be very low. For an introduction to EIT, we would recommend the article by S.E. Harris, *Electromagnetically* *induced transparency*, Physics Today (July 1997, pages 36–42). The article by C.L. Garrido Alzar, M.A.G. Martinez, and P. Nussensveig, *Classical analog of electromagnetically induced transparency*, Am. J. Phys. **70**, 37 (2001)[3] presents a detailed description of mechanical and electrical analogues of EIT.

---

[3] Also freely available online at http://arxiv.org/abs/quant-ph/0107061.

Chapter 4

Rotating Frames

4.1 The rotating saddle *

A particle of mass m can move in the plane (x, y) and experiences a conservative force corresponding to the potential:

$$U(x, y) = \frac{1}{2}m\omega^2(x^2 - y^2). \tag{4.1}$$

(1) Show that there exists one and only one equilibrium position for the particle. Is it stable?

(2) One rotates the system around the z axis, with angular velocity Ω. Write down the equations of motion in the rotating frame. For which values of Ω can the particle be trapped? Plot the stability diagram in the plane (ω, Ω).

(3) Figure 4.1(a) shows an experimental realization of such a rotating saddle. The equation of the surface reads $z = (x^2 - y^2)/R$, with $R = 0.45$ m, and the saddle is rotated at frequency $\nu = \Omega/(2\pi)$ by means of an electric motor. At time $t = 0$, one places a steel ball in the center of the saddle, and one measures the time τ after which the ball leaves the rotating saddle. Figure 4.1(b) shows the "lifetime" τ of the ball on the rotating saddle, as a function of ν. Comment. Are the experimental results in agreement with the theoretical prediction?

Solution

(1) The equilibrium positions correspond to stationary points of the potential (points where the partial derivatives vanish). Only the origin $(0,0)$

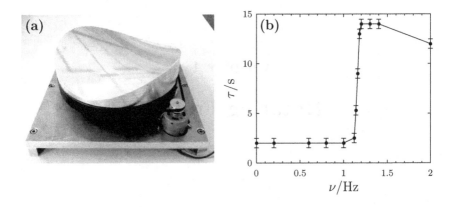

Fig. 4.1 (a) *Rotating saddle; one can see on the right the electric motor and the belt allowing the saddle to rotate.* (b) *Lifetime τ as a function of the rotation frequency ν. The experiment was performed by Dr. Tobias Koch (Stuttgart University).*

of the coordinate system is solution of the equations

$$\frac{\partial U}{\partial x} = \frac{\partial U}{\partial y} = 0.$$

It is therefore the only equilibrium position. It is obviously an *unstable* one, as it is not a local minimum of the potential, but a *saddle point*: the motion along y is unstable because

$$\frac{\partial^2 U}{\partial y^2} < 0.$$

(2) The lab frame being supposed inertial, one must add the centrifugal force $\boldsymbol{F}_{\text{cen}} = m\Omega^2 \boldsymbol{r}$ and the Coriolis force $-2m\boldsymbol{\Omega} \times \boldsymbol{v}$ (see the reminder at the end of the solution) when one writes down the equations of motion in the rotating frame:

$$m\ddot{\boldsymbol{r}} = -\boldsymbol{\nabla} U + m\Omega^2 \boldsymbol{r} - 2m\boldsymbol{\Omega} \times \dot{\boldsymbol{r}}.$$

By projecting on (x, y) one gets:

$$\ddot{x} - 2\Omega\dot{y} + \left(\omega^2 - \Omega^2\right) x = 0,$$
$$\ddot{y} + 2\Omega\dot{x} - \left(\omega^2 + \Omega^2\right) y = 0.$$

In order to deduce the stability of the solutions of this set of coupled differential equations, one looks for solutions of the form $x(t) = x_0 \exp(rt)$, $y(t) = y_0 \exp(rt)$, which leads to:

$$\left(r^2 + \omega^2 - \Omega^2\right) x_0 - 2r\Omega y_0 = 0,$$
$$2r\Omega x_0 + \left(r^2 - \omega^2 - \Omega^2\right) y_0 = 0.$$

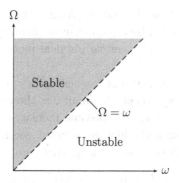

Fig. 4.2 *Stability diagram of the rotating saddle in the (ω, Ω) plane.*

This set of equations has solutions that differ from the trivial one $(x_0, y_0) = (0, 0)$ if and only if its determinant vanishes:

$$r^4 + 2\Omega^2 r^2 + \Omega^4 - \omega^4 = 0.$$

We need to solve a quadratic equation with unknown r^2. The discriminant is $\Delta = 4\omega^4$, and the solutions are therefore:

$$r^2 = -\Omega^2 \pm \omega^2.$$

If r is a root, $-r$ also, and consequently, if r is real, the particle motion diverges exponentially. The motion is bounded (i.e. the particle is trapped) if and only if one has oscillating solutions, thus if and only if r is purely imaginary, i.e. r^2 is real and negative. The stability condition therefore reads:

$$-\Omega^2 + \omega^2 < 0,$$

or, equivalently,

$$\Omega > \omega.$$

The stability diagram is shown in Fig. 4.2. One recovers the fact that for $\Omega = 0$, one has an instability, whatever ω. One needs to make the saddle rotate fast enough so that the Coriolis force stabilizes the motion (the centrifugal force cannot be the reason for this stabilization, as it tends to move the particle away from the center).

Traps for charged particles (the so-called Paul traps, extensively used in fundamental and applied physics nowadays) have a similar working principle: one makes an *a priori* unstable potential (in the static case) stable by making it time-dependent. An illustration of this phenomenon is given in Problem 8.4 by the quadrupole mass spectrometer; one will also find in Problem 8.6 a simple physical picture of the stabilization in the case $\Omega \gg \omega$.

(3) One observes that for rotation frequencies $\nu < \nu_0 = 1.15$ Hz, the ball leaves the saddle very quickly, in about 2 s: the motion is clearly unstable. However, for $\nu > \nu_0$, the lifetime increases abruptly: the motion is stabilized (although the lifetime is finite, see remark (iii) below). If one models the ball by a point-like particle, its potential energy reads

$$U = mgz = \frac{mg}{R}(x^2 - y^2). \tag{4.2}$$

Let us assume that the coordinate z of the ball is small, as well as its time-derivative $\dot{z}$, and that we can therefore neglect them. The dynamics is then limited to the plane (x, y), with a potential energy, Eq. (4.2), which corresponds to Eq. (4.1) with $\omega = \sqrt{2g/R}$, or numerically, $\omega/(2\pi) \simeq 1.05$ Hz. This result is not very different from the value ν_0 found experimentally. One therefore has a good quantitative agreement with theory, all the more given the crude approximations we have made.

♦ **Remarks.**

(i) Modeling the ball by a point-like particle is a very strong approximation. For instance, for a ball rolling, *without slipping*, on a plane making an angle α with respect to a horizontal plane, the reader will show that the acceleration of the center of mass of the ball is $7g \sin \alpha/9$, and is thus smaller than the acceleration $g \sin \alpha$ of a *sliding* point-like particle. This is due to the fact that the potential energy in the gravity field is converted, during the motion, into kinetic energy of the center of mass *and* into *rotational* kinetic energy!

(ii) The rotating saddle experiment can be described by the two-dimensional model used here only within the approximation $\dot{z} \ll \dot{x}, \dot{y}$. This is valid only if the ball remains close from the origin $(0, 0)$. Otherwise, motion along x, for instance, is not harmonic. One will study Problem 2.3 to understand what happens if one explicitly takes into account the velocity $\dot{z}$ of a particle moving, in the gravity field $-g\boldsymbol{u}_z$, on a curve $z(x)$.

(iii) One may wonder why the ball's "lifetime" is finite even for $\nu > \nu_0$. Besides the possible imperfections of the experimental setup, friction can explain this limited stability. Indeed, for a point-like particle, it is easy to convince oneself that a viscous friction force, for instance, makes the motion always unstable. Experimentally, if one reduces friction by lubricating the saddle, one indeed observes a strong increase of the ball's lifetime.

Reminder: Inertial forces

The acceleration $\boldsymbol{a}$ of a particle of mass m experiencing a force $\boldsymbol{F}$, in an inertial frame $\mathcal{R}$, fulfills Newton's second law:
$$\boldsymbol{F} = m\boldsymbol{a}.$$
The acceleration $\boldsymbol{a}'$ in a (non-inertial) frame $\mathcal{R}'$ moving with the angular velocity of rotation $\boldsymbol{\Omega}$ with respect to $\mathcal{R}$ is linked to $\boldsymbol{a}$ by
$$\boldsymbol{a}' = \boldsymbol{a} - 2\boldsymbol{\Omega} \times \boldsymbol{v}' - \boldsymbol{\Omega} \times (\boldsymbol{\Omega} \times \boldsymbol{r}')$$
with $\boldsymbol{v}'$ the velocity in $\mathcal{R}'$. For simplicity it is assumed here that the origins of $\mathcal{R}$ and $\mathcal{R}'$ coincide, and that $\boldsymbol{\Omega}$ is constant. This relationship comes only from kinematic considerations (see e.g. H. Goldstein, C. Poole and J. Safko, *Classical Mechanics*, Addison-Wesley, 2002, §4.9).

One can thus introduce fictitious *inertial forces*:

- the centrifugal force $\boldsymbol{F}_{\text{cen}} = -m\boldsymbol{\Omega} \times (\boldsymbol{\Omega} \times \boldsymbol{r}')$,
- the Coriolis force $\boldsymbol{F}_{\text{Cor}} = -2m\boldsymbol{\Omega} \times \boldsymbol{v}'$,

and write down Newton's equation in $\mathcal{R}'$ formally in the same way as for the case of an inertial frame:
$$m\boldsymbol{a}' = \boldsymbol{F} + \boldsymbol{F}_{\text{cen}} + \boldsymbol{F}_{\text{Cor}}.$$
For the case of a rotation around z, one easily sees that the expression for the centrifugal force reduces, in the plane (x, y), to the expression $m\Omega^2 \boldsymbol{r}'$ used in the solution above.

4.2 When Coriolis enters the game *

In this exercise, we propose, in a particular example, a direct derivation of the forces that need to be added to the external forces that are applied on a particle when the motion is described in a rotating frame (with respect to a given inertial frame), and to interpret their various effects.

We consider a particle of mass m that evolves in a two-dimensional space. The coordinates of this particle in the lab frame are (X, Y), and we assume that it is subjected to two kinds of forces, one of which depends explicitly on time. The equations of motion of this particle read:

$$\begin{cases} \ddot{X} = -\omega_0^2 X + \epsilon\omega_0^2(X\cos 2\Omega t - Y\sin 2\Omega t), & (4.3) \\ \ddot{Y} = -\omega_0^2 Y - \epsilon\omega_0^2(X\sin 2\Omega t + Y\cos 2\Omega t). & (4.4) \end{cases}$$

(1) Comment on these differential equations.
(2) To work out the physical content of the time-dependent force, we propose to use the following variables:

$$\begin{cases} x(t) = X(t)\cos\Omega t - Y(t)\sin\Omega t, \\ y(t) = X(t)\sin\Omega t + Y(t)\cos\Omega t. \end{cases} \quad (4.5)$$

What does this change of variables correspond to? Deduce the new expression of the equations of motion with the variables (x, y). Show that these variables allow one to work with effective forces that are time-independent.

(3) Interpret the three forces that appear in the equations of motion written with the new variables (x, y).

(4) We will address in the following the stability of the motion.

     (a) Discuss qualitatively the role of the different contributions of the forces, and the reason why one can expect instabilities.

     (b) To determine the stability domain, we will combine the two equations on x and y using the complex variable $z = x + iy$. Write down the equation of motion for the complex variable z.

     (c) Search for a solution in the form:

$$z = Ae^{i\lambda t} + Be^{-i\lambda t}, \qquad (A, B) \in \mathbb{R}^2.$$

Deduce the existence of an instability domain. Interpret.

Reminder. We recall the following trigonometric formulæ:

$$\sin(x)\sin(2x) + \cos(x)\cos(2x) = \cos(x),$$
$$\cos(x)\sin(2x) - \sin(x)\cos(2x) = \sin(x).$$

Solution

(1) There are clearly two types of forces. One is time-independent and has components $F_X = -m\omega_0^2 X$ and $F_Y = -m\omega_0^2 Y$. It corresponds to a harmonic restoring force towards the center of the lab frame. The other depends on time. However, we cannot develop further at this stage; we can only notice that this force depends linearly on the variables (X, Y) as would a harmonic force, but with eigenaxes that do not coincide with those of the inertial frame, and with a time-dependent angular frequency.

(2) The transformation from the inertial frame where the coordinates are labeled (X, Y) to the one in which the coordinates are labeled with (x, y) corresponds to going to a frame rotating with angular frequency $-\Omega$ (see Fig. 4.3).

In order to find the equations of motion fulfilled by (x, y), we first express (X, Y) as a function of (x, y) by solving Eq. (4.5):

$$\begin{cases} X(t) = x(t)\cos\Omega t + y(t)\sin\Omega t, \\ Y(t) = -x(t)\sin\Omega t + y(t)\cos\Omega t. \end{cases}$$

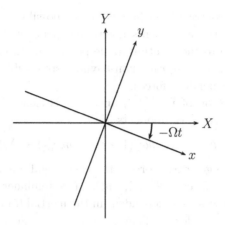

Fig. 4.3 *Representation of the axes (X, Y) of the lab frame and (x, y) of the rotating frame.*

Differentiating twice with respect to time, we find

$$\begin{cases} \ddot{X} = C\ddot{x} + S\ddot{y} - 2S\Omega\dot{x} + 2C\Omega\dot{y} - C\Omega^2 x - S\Omega^2 y \\ \ddot{Y} = -S\ddot{x} + C\ddot{y} - 2C\Omega\dot{x} - 2S\Omega\dot{y} + S\Omega^2 x - C\Omega^2 y, \end{cases}$$

where we have written, for short, $C = \cos\Omega t$ and $S = \sin\Omega t$. Replacing in Eq. (4.3) $\ddot{X}, X$ and Y by their expression as a function of (x, y) and their derivatives, we find

$$\left[\ddot{y} - 2\Omega\dot{x} - \Omega^2 y + \omega_0^2(1 + \epsilon)y\right]\cos\Omega t + $$
$$\left[-\ddot{x} - 2\Omega\dot{y} + \Omega^2 x + \omega_0^2(1 - \epsilon)x\right]\sin\Omega t = 0.$$

For this equality to hold at all times, the coefficients of $\cos\Omega t$ and $\sin\Omega t$ must both vanish, giving two equations fulfilled by (x, y). One checks in the same way that the equations obtained by the substitution in Eq. (4.4) are identical[1].

In fine, the equations motion in the rotating frame are:

$$\ddot{x} = -\omega_0^2\left(1 - \epsilon - \frac{\Omega^2}{\omega_0^2}\right)x - 2\Omega\dot{y}, \qquad (4.6)$$

$$\ddot{y} = -\omega_0^2\left(1 + \epsilon - \frac{\Omega^2}{\omega_0^2}\right)y + 2\Omega\dot{x}. \qquad (4.7)$$

The transformation we have carried out permits one to deal with second-order linear differential equations with time-independent coefficients. It is always possible to find an analytic solution of such a set of

---

[1]This is not obvious *a priori*, and shows that Eqs (4.3) and (4.4) actually have a very specific form.

equations. In contrast, it is in general not possible to find the explicit solution when the coefficients are time-dependent. In our case, the transformation to the rotating frame permits us to work out the analytical solution of an *a priori* non-trivial system subjected to a specific form of time-dependent forces.

(3) The interpretation of Eqs (4.6) and (4.7) is clear. We recognize the external force whose components are

$$\boldsymbol{F}_{\text{ext}} = \left(-m\omega_0^2(1 - \epsilon)x, -m\omega_0^2(1 + \epsilon)y\right).$$

In addition, we get extra forces: the centrifugal force $(m\Omega^2 x, m\Omega^2 y)$, and the Coriolis force $(-2m\Omega\dot{y}, 2m\Omega\dot{x})$ (see reminder on page 49).

The force that was time-dependent in the inertial frame adds up to the harmonic restoring force. They give rise to a time-independent force that corresponds to an anisotropic harmonic force in two dimensions with the following potential energy U_{ext}:

$$\boldsymbol{F}_{\text{ext}} = -\boldsymbol{\nabla}U_{\text{ext}} \text{ with } U_{\text{ext}} = \frac{1}{2}m\omega_0^2(1 - \epsilon)x^2 + \frac{1}{2}m\omega_0^2(1 + \epsilon)y^2.$$

In the inertial frame, the eigenaxes of this harmonic potential are rotating around the center O at an angular velocity $-\Omega$. The centrifugal force $\boldsymbol{F}_{\text{cen}}$ tends to decrease the strength of the harmonic restoring force since it corresponds to an expelling force, as intuitively expected. It is also associated with a potential:

$$\boldsymbol{F}_{\text{cen}} = -\boldsymbol{\nabla}U_{\text{cen}} \text{ with } U_{\text{cen}} = -\frac{1}{2}m\Omega^2(x^2 + y^2).$$

The Coriolis force $\boldsymbol{F}_{\text{cor}}$ cannot be expressed as the gradient of a potential. Formally, this force is analogous to the Lorentz force which acts on a charged particle in the presence of a magnetic field. This force cannot change the modulus of the velocity, but tends to curve the trajectories. Its expression is given by

$$\boldsymbol{F}_{\text{cor}} = -2m\boldsymbol{\Omega} \times \boldsymbol{v},$$

where the components of the rotation vector are $\boldsymbol{\Omega} = (0, 0, -\Omega)$. These coordinates are expressed in three dimensions since the rotation vector is perpendicular to the plane in which the trajectory lies. Its expression is the same in the lab and rotating frames.

(4) (a) Qualitatively, one could say that the centrifugal force tends to destabilize the motion. Indeed, when $\Omega^2 \geqslant \omega_0^2(1 \pm \epsilon)$, the harmonic confinement does not suffice to counteract the centrifugal force. In the absence of the Coriolis force, one could conclude that

the motion would be unstable as soon as $\Omega^2 > \omega_0^2(1 - \epsilon)$ since the differential equation for x would take the simple form

$$\ddot{x} = \tilde{\omega}^2 x \text{ with } \tilde{\omega}^2 = \Omega^2 - \omega_0^2(1 - \epsilon) > 0,$$

and the particle would be expelled from the center O. The next questions are devoted to clarifying the role played by the Coriolis force in this context.

(b) We directly obtain:

$$\ddot{z} - 2i\Omega\dot{z} + (\omega_0^2 - \Omega^2)z - \omega_0^2\epsilon\bar{z} = 0, \tag{4.8}$$

where $\bar{z} = x - iy$ is the complex conjugate of z.

(c) Substituting the proposed solution into Eq. (4.8), we find

$$[-\lambda^2 A + (\omega_0^2 - \Omega^2)A - \omega_0^2\epsilon B + 2\Omega\lambda A]e^{i\lambda t}$$
$$+ [-\lambda^2 B + (\omega_0^2 - \Omega^2)B - \omega_0^2\epsilon A - 2\Omega\lambda B]e^{-i\lambda t} = 0.$$

This relation has to be valid at any time t; thus the coefficients in between the brackets necessarily vanish:

$$\begin{bmatrix} \lambda^2 - \omega_0^2 + \Omega^2 - 2\Omega\lambda & \omega_0^2\epsilon \\ \omega_0^2\epsilon & \lambda^2 - \omega_0^2 + \Omega^2 + 2\Omega\lambda \end{bmatrix} \begin{bmatrix} A \\ B \end{bmatrix} = 0.$$

This set of linear equations possesses non-trivial solutions $(A, B) \neq (0, 0)$ if and only if the determinant is zero:

$$\lambda^4 - 2\lambda^2(\omega_0^2 + \Omega^2) + (\omega_0^2 - \Omega^2)^2 - \omega_0^4\epsilon^2 = 0.$$

We get a quadratic equation in the variable λ^2, whose solutions are:

$$\lambda_{\pm}^2 = \omega_0^2 + \Omega^2 \pm \omega_0\sqrt{4\Omega^2 + \epsilon^2\omega_0^2}.$$

The oscillating solutions that correspond therefore to stable trajectories fulfill the inequality $\lambda_{\pm}^2 > 0$. They imply that the the rotation angular velocity Ω belongs either to the interval $[0, \omega_0\sqrt{1 - \epsilon}[$, or to $]\omega_0\sqrt{1 + \epsilon}, \infty[$. In the first interval, the stability makes sense since the centrifugal force is not sufficient to counteract the harmonic restoring force. The results are much more surprising for the second interval since the centrifugal force overcomes the restoring force. The Coriolis force plays a key role in explaining the stability in this region of parameters. It is one of the few examples for which the Coriolis force really controls the dynamics of the particle[2]. The Coriolis force makes the velocity vector rotate, or, to put

---

[2]Two other examples of dynamical stabilization are studied in Problems 4.1 (The rotating saddle) and 10.2 (Lagrange points).

Lab frame   Rotating frame

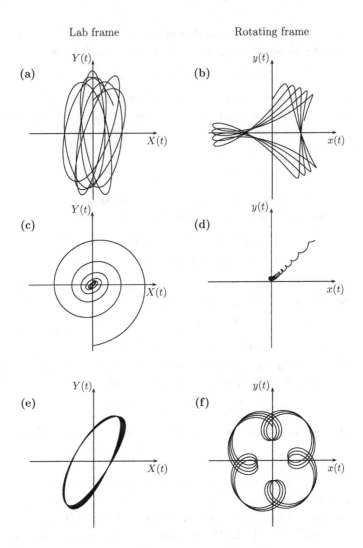

Fig. 4.4 *Numerical resolution of the equations of motion with $\epsilon = 0.2$, and the following initial conditions in the inertial frame $X(0) = 0$, $\dot{X}(0) = 1$, $Y(0) = 1$ and $\dot{Y}(0) = 1$. (a) and (b) $\Omega = 0.3\,\omega_0$, (c) and (d) $\Omega = \omega_0$, (e) and (f) $\Omega = 1.5\,\omega_0$. The trajectory remains bounded for cases (a), (b), (e), (f), and diverges for (c) and (d), as expected from our calculations.*

it another way, the velocity vector precesses around the direction of the rotation vector. If the velocity vector rotates sufficiently fast, the expelling force that results from the external force combined to the centrifugal force is counteracted because of the rotation of the velocity vector. Such a situation is referred to as a *dynamical stabilization*. In the instability domain $\Omega/\omega_0 \in \left[\sqrt{1-\epsilon}, \sqrt{1+\epsilon}\right]$, the Coriolis force does not make the velocity vector rotate sufficiently fast to compensate for the expelling force, and the particle escapes exponentially from the origin (see Fig. 4.4).

Chapter 5

Oscillators

5.1 Oscillation period **

In this exercise, we consider only a one-dimensional situation.

(1) Figure 5.1 represents an even potential well $U(x)$ as a function of the position x. We consider a particle of energy E oscillating in this well. Establish the expression for the density distribution of time presence per unit length for the particle in the potential $U(x)$. Deduce the general expression for the oscillation period. Calculate its explicit expression for a harmonic potential $U(x) = m\omega^2 x^2/2$.

(2) We consider a particle of mass m with an energy E that experiences the potential $U(x) = U_0 \left[1 - \cos(x/a)\right]$. Find the condition on E and U_0 for which the particle remains trapped in one of the potential wells

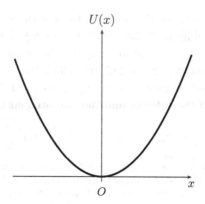

Fig. 5.1 *A symmetric potential well.*

Table 5.1 *Measured period T of a simple pendulum of length*
$\ell \simeq 1.00$ m, *as a function of the amplitude θ_0 of the oscillations.*

θ_0 (°)	10	20	30	40	50	60	70	80
T (s)	2.00	2.02	2.04	2.07	2.10	2.15	2.21	2.27

of $U(x)$. Assuming that this relation is fulfilled, calculate, for the well centered at $x = 0$, the oscillation period $T(E)$. The result will be expressed as a power series in the energy E. We recall the following expansion, valid for $|x| < 1$:

$$\frac{1}{(1-x)^{1/2}} = \sum_{n=0}^{\infty} \frac{(2n)!}{2^{2n}(n!)^2} x^n. \tag{5.1}$$

(3) **Application: Borda's formula.** Show that the oscillation period of a simple pendulum of length ℓ can be expanded as a function of the oscillation amplitude θ_0 as

$$T = 2\pi \sqrt{\frac{\ell}{g}} \left(1 + \frac{\theta_0^2}{16} \right). \tag{5.2}$$

This result is known as Borda's formula. Table 5.1 shows a set of measurements of the period[1] $T(\theta_0)$ for a pendulum of length $\ell \simeq 1.00$ m. Is Borda's formula valid for these experimental points?

Solution

(1) For a given energy E of the particle, the positions $\pm x_0$ of the turning points are the solution of the equation $U(\pm x_0) = E$ [see Fig. 5.2(a)]. Let us denote by $P(x)$ the probability density of time presence per unit length of the particle. By definition, $P(x)|dx| = Adt$, where A is a constant. The conservation of energy gives the relation between the absolute value of the velocity and the potential energy:

$$E = \frac{1}{2}m\dot{x}^2 + U(x) \qquad \Longrightarrow \qquad \left|\frac{dx}{dt}\right| = \sqrt{\frac{2}{m}[E - U(x)]}.$$

---

[1]The reader can easily perform his/her own set of measurements. One just needs to attach a massive bob at the end of a string of length ℓ. Choosing $\ell \sim 1$ m yields a period on the order of $T \sim 2$ s, which can be measured easily. More accurate measurements of the period can be carried out by counting many periods; however, one has to check that the amplitude of the oscillations does not decrease significantly during the measurement because of the unavoidable residual friction.

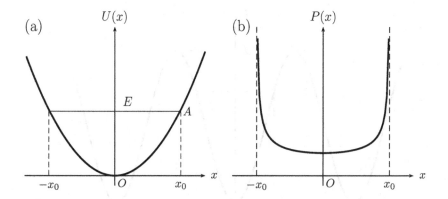

Fig. 5.2 (a) *Potential energy* $U(x)$. (b) *Probability density* $P(x)$ *of time presence per unit length of the particle in the potential well.*

We deduce the expression for $P(x)$:

$$P(x) = \frac{A'}{\sqrt{E - U(x)}},$$

where A' is a constant. This probability density diverges at the turning points $x = \pm x_0$ [see Fig. 5.2(b)]. This is caused by the decrease of the velocity when the particle approaches the turning points. The velocity vanishes at the turning points and changes sign.

A quarter of the oscillation period corresponds to the path from O to A [see Fig. 5.2(a)]. We thus deduce the expression for the oscillation period for an even potential well:

$$\frac{T}{4} = \int_0^{T/4} dt = \int_0^{x_0} \frac{dx}{v(x)} = \int_0^{x_0} \sqrt{\frac{m}{2}} \frac{dx}{\sqrt{E - U(x)}}. \tag{5.3}$$

The application to the harmonic oscillator $U(x) = m\omega^2 x^2/2$ is straightforward:

$$\frac{T}{4} = \frac{1}{\omega} \int_0^{x_0} \frac{dx}{\sqrt{x_0^2 - x^2}} = \frac{1}{\omega} \int_0^1 \frac{du}{\sqrt{1 - u^2}} = \frac{1}{\omega} \int_0^{\pi/2} d\varphi,$$

where we have successively performed the following changes of variables $u = x/x_0$, and $\varphi = \arcsin u$. In this way, we recover the well-known result $T = 2\pi/\omega$. For a harmonic potential, the oscillation period does not depend on the energy of the particle. This property is referred to as the *isochronism* of the harmonic oscillator. It is not valid for an arbitrary potential well, as illustrated in the second part of the exercise.

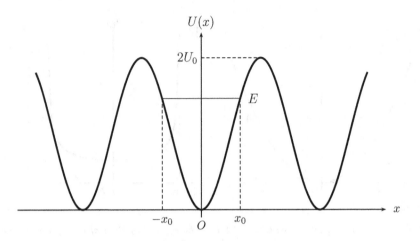

Fig. 5.3   *A periodic potential. If the particle energy E is less than the individual potential well depth $E < 2U_0$, the particle remains in one well only and oscillates between $-x_0$ and x_0: one has a bound state.*

(2) The particle is confined in a potential well if and only if $E < 2U_0$ (see Fig. 5.3). In this case, we define as above the turning points for the well centered at $x = 0$ by $\pm x_0$, solutions of the equation $U(\pm x_0) = E$. Using Eq. (5.3), the quarter of the period reads:

$$\frac{T}{4} = \left(\frac{m}{2U_0}\right)^{1/2} \int_0^{x_0} \frac{dx}{[\cos(x/a) - \cos(x_0/a)]^{1/2}}.$$

Using the change of variables $\sin\psi = \sin[x/(2a)]/\sin[x_0/(2a)]$, we get

$$\frac{T}{4} = \left(\frac{m}{U_0}\right)^{1/2} \int_0^{\pi/2} \frac{d\psi}{\left[1 - \sin^2[x_0/(2a)]\sin^2\psi\right]^{1/2}}.$$

The solution is therefore expressed through an *elliptical integral*[2], that cannot be recast in terms of elementary functions. This is the reason why we perform a series expansion using Eq. (5.1). We find:

$$T = 4\left(\frac{m}{U_0}\right)^{1/2} \sum_{n=0}^{\infty} \frac{(2n)!}{2^{2n}(n!)^2}\left(\sin^2\frac{x_0}{2a}\right)^n I_n,$$

where I_n are the Wallis' integrals defined by

$$I_n = \int_0^{\pi/2} \sin^{2n}\psi\, d\psi.$$

---

[2]More precisely, this is the so-called *complete elliptic integral of the first kind.*

One can easily show, using integration by parts, the following relation $2nI_n = (2n - 1)I_{n-1}$, valid for $n \geqslant 1$. Obviously, $I_0 = \pi/2$, so that one has

$$I_n = \frac{(2n)!}{2^{2n}(n!)^2} \frac{\pi}{2}.$$

We finally deduce the exact expression for the oscillation period as a series expansion in the energy E:

$$T(E) = 2\pi \left(\frac{ma^2}{U_0}\right)^{1/2} \sum_{n=0}^{\infty} \left[\frac{(2n)!}{2^{2n}(n!)^2}\right]^2 \left(\frac{E}{2U_0}\right)^n.$$

In contrast with the harmonic potential, the oscillation period depends explicitly on the energy E of the particle.

However, at very low energy ($E \ll 2U_0$), only the first terms of the expansion contribute to the series:

$$T(E) \simeq 2\pi \left(\frac{ma^2}{U_0}\right)^{1/2} \left(1 + \frac{E}{8U_0} + \dots\right). \tag{5.4}$$

The first term of this expansion does not depend on the energy since it corresponds to the domain of validity of the harmonic expansion of the potential well $U(x)$ about its minimum:

$$U(x) \simeq U_0 \frac{x^2}{2a^2} \equiv \frac{1}{2}m\omega^2 x^2,$$

where we have introduced the angular frequency $\omega = \left(U_0/(ma^2)\right)^{1/2}$. The first term of the expansion of the oscillation period in Eq. (5.4) indeed corresponds to the expected result for a harmonic potential $T \simeq 2\pi/\omega \equiv T_0$. The second term $E/(8U_0)$ of the expression (5.4) for the oscillation period is the first correction to the period T_0 and is caused by the quartic term (i.e. the term proportional to x^4) of the potential that arises from the expansion of the potential energy around $x = 0$. When the energy E approaches the threshold $2U_0$, the accurate evaluation of the oscillation period requires that we take into account an increasingly large number of terms in the expansion of $T(E)$. The value of the oscillation period T increases as compared to the one obtained for the harmonic approximation; when E approaches $2U_0$, the oscillation period diverges: $T(E) \to \infty$.

(3) **Application: Borda's formula.** A simple pendulum corresponds to a sinusoidal potential energy $U(\theta) = mg\ell(1 - \cos\theta)$, whence $U_0 = mg\ell$.

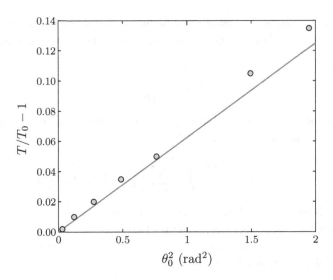

Fig. 5.4   *Relative difference $T/T_0 - 1$ between the oscillation period T of a simple pendulum and its value T_0 for a vanishingly small amplitude (harmonic approximation),
as a function of θ_0^2, where θ_0 is the oscillation amplitude. Borda's formula, Eq. (5.2),
corresponds to a straight line with a slope 1/16 (gray line).*

According to Eq. (5.4), the first correction to the harmonic approximation of the oscillation period, $T_0 = 2\pi\sqrt{\ell/g}$, is therefore given by

$$\frac{E}{8U_0} = \frac{mg\ell(1 - \cos\theta_0)}{8mg\ell} \simeq \frac{\theta_0^2}{16},$$

which proves Borda's formula. This first correction is relatively small,
since it yields a modification of the period of only 6% for $\theta_0 = 1$, that
is for an oscillation amplitude equal to $\sim 57°$. This is the reason why
Galileo concluded, wrongly, that the simple pendulum is isochronous.
To check Borda's formula with the experimental data of Table 5.1, we
plot the relative difference $T/T_0 - 1$ as a function of θ_0^2. For small
angular amplitudes $\theta_0 \ll 1$, the experimental points are in very good
agreement with Borda's formula, which corresponds here to a straight
line of slope 1/16, as illustrated in Fig. 5.4.

5.2   Optimal transport of a bunch of atoms **

The goal of this problem is to show, in a simple, one-dimensional model, that a bunch of trapped atoms can be moved by displacing its confining potential, without undergoing any excitation of the center of mass (for given conditions to be determined). The potential well is displaced from a point O_1, located at $x = -d$, to the point O_2 at $x = 0$ (see Fig. 5.5).

The minimum of the harmonic potential has a position noted $\delta(t)$ in the (inertial) lab frame. The potential undergoes a constant acceleration a for a duration τ, and then a constant deceleration $-a$ over the same time τ (see Fig. 5.6).

(1) Give the expression of d and of v_0 (the maximal velocity during the transport) as a function of τ and a.
(2) Give the expression of $\delta(t)$ for the two time intervals $0 \leq t \leq \tau$ and $\tau \leq t \leq 2\tau$.
(3) The equation of motion for an atom in the moving harmonic potential is simply:

$$\frac{\mathrm{d}^2 x}{\mathrm{d}t^2} + \omega_0^2 \left[x - \delta(t) \right] = 0,$$

where ω_0 is the angular frequency of the harmonic potential. For the sake of simplicity, in the following, we normalize time to the trap frequency by introducing the new variable $\hat{t} = \omega_0 t$. We denote by $\boldsymbol{X}$ the vector $(x, \mathrm{d}x/\mathrm{d}\hat{t})$.

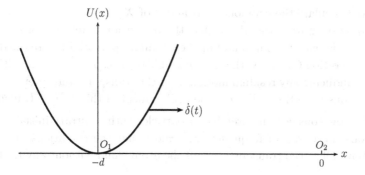

Fig. 5.5   *A harmonic potential well is moved from $x = -d$ to $x = 0$, allowing the transport of the atoms confined in it.*

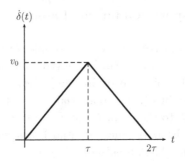

Fig. 5.6   *Velocity profile $\dot{\delta}(t)$ of the potential well.*

Show that

$$\frac{\mathrm{d}\boldsymbol{X}}{\mathrm{d}\hat{t}} = \mathsf{A}\boldsymbol{X} + \boldsymbol{B}(\hat{t}), \qquad (5.5)$$

where A is a 2×2 matrix and $\boldsymbol{B}(\hat{t})$ a vector. Give the expressions of A and $\boldsymbol{B}$.

(4) One looks for a solution of Eq. (5.5) having the form:

$$\boldsymbol{X}(\hat{t}) = \mathrm{e}^{\mathsf{A}\hat{t}}\boldsymbol{X}_0(\hat{t}),$$

where e^{M} denotes the exponential of the matrix M (see the reminder at the end of the solution). Write down the formal expression of $\boldsymbol{X}_0(\hat{t})$ as a function of $\boldsymbol{X}_0(0)$, A and $\boldsymbol{B}$.

(5) Calculate $\mathrm{e}^{\mathsf{A}\hat{t}}$. Comment.

(6) One first solves the equations of motion for a particle that was initially at rest at the bottom of the well.

   (a) Calculate the components α and β of $\boldsymbol{X}_0(2\tau)$.

   (b) Deduce from the above that the oscillation amplitude of the particle once the minimum of the confining potential has reached the position O_2. Show that it is possible to transport the atom without inducing any residual motion after the transport. Interpret the various conditions allowing the achievement of this type of transport.

(7) We now consider an ensemble of particles with arbitrary initial conditions, in a trap of frequency ω_0. Show that the conditions obtained above allow the transportation of the atom cloud without any heating.

Solution

(1) After an acceleration a over a time τ, the velocity is $v_0 = a\tau$. The distance that has been covered at τ is $a\tau^2/2$. The total distance covered by the trap after the acceleration and deceleration phases is thus $d = a\tau^2$.

(2) We know the velocity of the trap for all times, and its initial position, we thus deduce immediately, upon integration, its position:

$$0 \le t \le \tau: \qquad\qquad \delta(t) = -d + at^2/2,$$
$$\tau \le t \le 2\tau: \quad \delta(t) = \delta(\tau) + a\tau(t - \tau) - a(t - \tau)^2/2.$$

(3) The equation of motion written in terms of the normalized time reads:

$$\frac{d^2 x}{d\hat{t}^2} + [x - \delta(\hat{t})] = 0.$$

We immediately deduce the evolution of the vector $\boldsymbol{X} = (x, dx/d\hat{t})$:

$$\frac{d\boldsymbol{X}}{d\hat{t}} = \mathsf{A}\boldsymbol{X} + \boldsymbol{B}(\hat{t}), \qquad\qquad (5.6)$$

with

$$\mathsf{A} = \begin{pmatrix} 0 & 1 \\ -1 & 0 \end{pmatrix} \qquad \text{and} \qquad \boldsymbol{B}(\hat{t}) = \begin{pmatrix} 0 \\ \delta(\hat{t}) \end{pmatrix}.$$

(4) By introducing the proposed solution $\boldsymbol{X}(\hat{t}) = e^{\mathsf{A}\hat{t}}\boldsymbol{X}_0(\hat{t})$ into Eq. (5.6), we deduce

$$\frac{d\boldsymbol{X}_0}{d\hat{t}} = e^{-\mathsf{A}\hat{t}}\boldsymbol{B}(\hat{t}),$$

and thus the formal expression of $\boldsymbol{X}_0(\hat{t})$:

$$\boldsymbol{X}_0(\hat{t}) = \boldsymbol{X}_0(0) + \int_0^{\hat{t}} e^{-\mathsf{A}\hat{\tau}'} \boldsymbol{B}(\hat{\tau}') \, d\hat{\tau}'. \qquad\qquad (5.7)$$

(5) In order to find the explicit solution, we have to calculate $e^{\mathsf{A}\hat{t}}$. We first diagonalize the matrix A, which can be written as $\mathsf{A} = \mathsf{R}\mathsf{D}\mathsf{R}^{-1}$ where D is a diagonal matrix, with coefficients that are the eigenvalues of A, and R is the change-of-basis matrix to the eigenbasis:

$$\mathsf{R} = \frac{1}{\sqrt{2}}\begin{pmatrix} 1 & i \\ i & 1 \end{pmatrix}, \quad \mathsf{D} = \begin{pmatrix} i & 0 \\ 0 & -i \end{pmatrix} \quad \text{and} \quad \mathsf{R}^{-1} = \frac{1}{\sqrt{2}}\begin{pmatrix} 1 & -i \\ -i & 1 \end{pmatrix}.$$

The exponential of the matrix is calculated using the series expansion:

$$e^{\mathsf{A}\hat{t}} = \sum_{n=0}^{\infty} \frac{(\mathsf{A}\hat{t})^n}{n!} = \sum_{n=0}^{\infty} \frac{(\mathsf{R}\mathsf{D}\mathsf{R}^{-1}\hat{t})^n}{n!} = \mathsf{R}\left(\sum_{n=0}^{\infty} \frac{(\mathsf{D}\hat{t})^n}{n!}\right)\mathsf{R}^{-1} = \mathsf{R}e^{\mathsf{D}\hat{t}}\mathsf{R}^{-1},$$

or, finally:

$$e^{A\hat{t}} = \frac{1}{2}\begin{pmatrix} 1 & i \\ i & 1 \end{pmatrix}\begin{pmatrix} e^{i\hat{t}} & 0 \\ 0 & e^{-i\hat{t}} \end{pmatrix}\begin{pmatrix} 1 & -i \\ -i & 1 \end{pmatrix} = \begin{pmatrix} \cos\hat{t} & \sin\hat{t} \\ -\sin\hat{t} & \cos\hat{t} \end{pmatrix}.$$

The matrix $e^{A\hat{t}}$ thus corresponds to a mere rotation in phase space.

(6) (a) Reporting into Eq. (5.7), we obtain the components of $X_0(2\hat{\tau})$:

$$\begin{pmatrix} \alpha \\ \beta \end{pmatrix} = \begin{pmatrix} -d \\ 0 \end{pmatrix} + \int_0^{2\hat{\tau}} \begin{pmatrix} \cos\hat{\tau}' & -\sin\hat{\tau}' \\ \sin\hat{\tau}' & \cos\hat{\tau}' \end{pmatrix}\begin{pmatrix} 0 \\ \delta(\hat{\tau}') \end{pmatrix} d\hat{\tau}'.$$

$$\implies \quad \begin{cases} \alpha = -d - \omega_0 \displaystyle\int_0^{2\tau} \delta(\tau')\sin(\omega_0\tau')\,d\tau', \\ \beta = \omega_0 \displaystyle\int_0^{2\tau} \delta(\tau')\cos(\omega_0\tau')\,d\tau' \end{cases} \tag{5.8}$$

To calculate these integrals, we take into account the explicit expression of $\delta(t)$ for the two time intervals to be considered. After a long but easy calculation, we get a very simple result:

$$\begin{cases} \alpha = \dfrac{2a}{\omega_0^2}\cos\left(\dfrac{v_0\omega_0}{a}\right)\left[\cos\left(\dfrac{v_0\omega_0}{a}\right) - 1\right], \\ \beta = \dfrac{2a}{\omega_0^2}\sin\left(\dfrac{v_0\omega_0}{a}\right)\left[\cos\left(\dfrac{v_0\omega_0}{a}\right) - 1\right]. \end{cases}$$

(b) From the form of the solution $X(\hat{t}) = e^{A\hat{t}}X_0(\hat{t})$, we directly deduce the expression of the position $x(2\tau)$ and of the velocity $\dot{x}(2\tau)$ at time 2τ:

$$\begin{cases} x(2\tau) = \alpha\cos[2\omega_0\tau] + \beta\sin[2\omega_0\tau], \\ \dot{x}(2\tau) = -\alpha\omega_0\sin[2\omega_0\tau] + \beta\omega_0\cos[2\omega_0\tau]. \end{cases}$$

Using energy conservation, we deduce (for $t > 2\tau$) the amplitude $\delta_{\max}$ of the oscillations when the trap is fixed at O_2:

$$\dot{x}^2(2\tau) + \omega_0^2 x^2(2\tau) = \omega_0^2\delta_{\max}^2 = \omega_0^2(\alpha^2 + \beta^2). \tag{5.9}$$

We finally obtain:

$$\delta_{\max}(a, v_0, \omega_0) = \frac{2a}{\omega_0^2}\left|\cos\left(\frac{v_0\omega_0}{a}\right) - 1\right| = \frac{4a}{\omega_0^2}\sin^2\left(\frac{v_0\omega_0}{2a}\right).$$

For given values of (a, v_0), we have plotted on Fig. 5.7 the variations of $\delta_{\max}$ versus the angular frequency ω_0 of the trap. The higher ω_0, i.e. the tighter the confinement, the smaller the amplitude of the excitations induced by the transport. On the contrary, in a weak trap (i.e. with a small ω_0), one has strong oscillations after

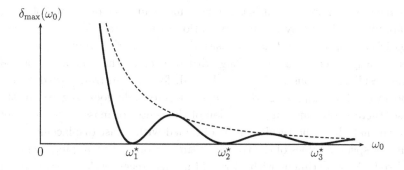

Fig. 5.7 *Oscillation amplitude $\delta_{max}(\omega_0)$ of an atom initially at rest, transported by displacing the harmonic trap that confines it. The dashed curve represents the amplitude of δ_{max}, and varies as $1/\omega_0^2$. The points $\omega_i^\star$ correspond to frequencies that make the oscillation amplitude vanish after transport, i.e. that allow for an optimal transport, without any excitation after the move.*

the transport, because the particle must explore the potential over a large distance to compensate for the acceleration.

For some "magic" values $\omega_p^\star$ of the trap angular frequency, the oscillation amplitude after transport vanishes, and the particle stands still at the bottom of the potential well:

$$\frac{v_0\omega_p^\star}{2a} = p\pi, \qquad p \in \mathbb{N}. \qquad (5.10)$$

These resonance conditions simply appear when the work done by the force associated with the moving potential during acceleration is compensated during deceleration.

(7) For an ensemble of particles, we need to solve the problem for arbitrary initial conditions. The only modification consists in changing α into $\alpha + \alpha_0$ and β into $\beta + \beta_0$, where the constants α_0 and β_0 take into account the initial conditions. Energy conservation then reads:

$$\mathcal{E} = \dot{x}^2(2\tau) + \omega_0^2 x^2(2\tau) = \omega_0^2(\alpha^2 + \beta^2 + \alpha_0^2 + \beta_0^2 - 2\alpha\alpha_0 - 2\beta\beta_0).$$

If the condition (5.10) of optimal transport is fulfilled, $\alpha = \beta = 0$, and $\mathcal{E}$ is equal to its initial value $\mathcal{E} = \alpha_0^2 + \beta_0^2$. The transport thus does not add any energy to the system. This is remarkable, since, in general, the energy is not any more a constant of motion in the presence of time-dependent forces. Condition (5.10) is thus also valid for the transport of an arbitrary atom cloud!

We have thus shown that it is possible to move an atom or a cloud of atoms "rapidly", without any residual excitation of its center of mass. An "adiabatic" transport would have consisted in moving the trapping potential very slowly, over a timescale long with respect to the oscillation period. This result is obvious in Fig. 5.7. Indeed, for a given transport duration, and in the limit $\omega_0 \to \infty$, the oscillation amplitude goes to zero. In this case, there is obviously no excitation of the center of mass. Here, we have shown that the same result can be obtained with a fast displacement, but only for specific values of the transport time 2τ once the trapping frequency is fixed. An experiment with ultracold atoms[3] recently demonstrated this technique of non-adiabatic transport.

Let us mention another important point. Equation (5.8) shows that the residual oscillation amplitude is directly related to the Fourier transform of the trap position $\delta(t)$. This remark allows for interesting analogies with optics. Indeed, Fraunhofer diffraction gives an intensity profile which is just the square of the Fourier transform of the transmission $t(x)$ of the diffracting object. Looking for the "magic" durations for the optimal transport thus amounts to finding the dark fringes in the corresponding diffraction pattern. We can thus relate a mechanics problem to an optics one. The well known techniques developed in the latter field (such as apodization for instance) can then be used in this transport problem!

Reminder: Matrix exponentiation

The exponential e^M of the (real or complex) matrix M is defined by the power series

$$e^M = \exp M = \sum_{k=0}^{\infty} \frac{1}{k!} M^k.$$

This series is always convergent. Note that, in general, $\exp(A + B) \neq \exp A \exp B$. However the equality holds if A and B commute.

It is obvious from the definition that the exponential of a diagonal matrix $\mathrm{diag}(a_1, \ldots, a_N)$ is also a diagonal matrix given by $\mathrm{diag}(\exp a_1, \ldots, \exp a_N)$. Therefore, in order to compute the exponential of a diagonalizable matrix, one usually diagonalizes it first, as in the exercise above.

---

[3] A. Couvert, T. Kawalec, G. Reinaudi, and D. Guéry-Odelin, *Optimal transport of ultracold atoms in the non-adiabatic regime*, Europhys. Lett. **83**, 13001 (2008), also available online at http://arxiv.org/abs/0708.4197.

5.3 Trajectory in a curved guide**

We consider a particle M, of mass m, moving in a curved guide. Transversally to the local guide axis, the particle is confined by a harmonic potential energy $U(y) = m\omega^2 y^2/2$ (see Fig. 5.8). The aim of this exercise is to determine the modifications of the trajectory that occur in the curved part of the guide. To simplify, we consider only one transverse dimension y.

The radius vector of the particle is parametrized by the arc length s through its orthogonal projection x on the local guide axis and its transverse coordinate y:

$$r = x(s) + yn(s).$$

We denote by t the tangent unit vector at $x(s)$, and by n the normal unit vector. We recall the well-known Frenet-Serret relations for a plane curve:

$$t = \frac{dx}{ds}, \qquad \frac{dt}{ds} = \kappa n, \qquad \frac{dn}{ds} = -\kappa t,$$

where κ^{-1} is the radius of curvature, that is non-zero only from A to B in our case.

(1) Give the expression for the components of the velocity v on t and n.
(2) Give the expression for the components of the acceleration a on t and n.
(3) Apply Newton's second law and deduce a relation between $u \equiv ds/dt$ and y.
(4) Solve the equation of motion along the y axis under the assumption $\kappa y \ll 1$. Comment.

Solution

(1) By definition of the velocity vector:

$$v = \frac{dr}{dt} = \frac{dx}{dt} + \frac{dy}{dt}n - \kappa y \frac{ds}{dt}t = \dot{s}(1 - \kappa y)t + \dot{y}n.$$

(2) Similarly, from the definition of the acceleration we deduce :

$$a = \frac{dv}{dt} = [\ddot{s}(1 - \kappa y) - 2\kappa \dot{s}\dot{y}]\,t + [\ddot{y} + \kappa(1 - \kappa y)\dot{s}^2]\,n.$$

(3) Here, Newton's second law reads:

$$ma = -\nabla U = -m\omega^2 yn,$$

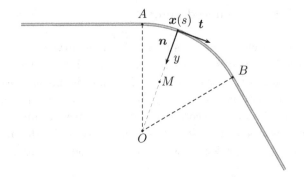

Fig. 5.8 *Guide axis (gray line) and notations. The position of the point M is exaggeratedly far from the guide to facilitate the representation of the tangent and normal unit vectors. In practice we consider a case for which the distance to the guide axis is very small compared to the radius of curvature.*

whence

$$\frac{\mathrm{d}^2 s}{\mathrm{d}t^2}(1 - \kappa y) - 2\kappa \frac{\mathrm{d}s}{\mathrm{d}t}\frac{\mathrm{d}y}{\mathrm{d}t} = 0$$

$$\frac{\mathrm{d}^2 y}{\mathrm{d}t^2} + \omega^2 y + \left(\frac{\mathrm{d}s}{\mathrm{d}t}\right)^2 (1 - \kappa y)\kappa = 0.$$

Let us introduce the quantity $u = \mathrm{d}s/\mathrm{d}t$. The first equation, assuming $\kappa y < 1$, can be recast in the form

$$\frac{\dot{u}}{u} = \frac{2\kappa \dot{y}}{1 - \kappa y},$$

which upon integration yields a relation between u and the transverse coordinate y:

$$\frac{u}{u_0} = \left(\frac{1 - \kappa y_0}{1 - \kappa y}\right)^2, \tag{5.11}$$

where u_0 is the longitudinal velocity of the particle at point A and y_0 its transverse position. Combining the previous relations, we finally obtain the following differential equation for the variable y:

$$\frac{\mathrm{d}^2 y}{\mathrm{d}t^2} + \omega^2 y + (1 - \kappa y_0)\kappa u_0^2 \left(\frac{1 - \kappa y_0}{1 - \kappa y}\right)^3 = 0. \tag{5.12}$$

This equation implies energy conservation. To show this, we multiply Eq. (5.12) by $\mathrm{d}y/\mathrm{d}t$ and we integrate over time. Using Eq. (5.11), we find:

$$\frac{1}{2}v^2 + \frac{1}{2}\omega^2 y^2 = \frac{1}{m}\left(\frac{1}{2}mv^2 + U(y)\right) = \text{const.}$$

In the limit $\kappa y \ll 1$, the effect of the curvature can be derived from an expansion of the third term of Eq. (5.12):

$$\frac{1}{(1 - \kappa y)^3} \simeq 1 + 3\kappa y.$$

To the lowest order in κy, we find

$$(1 - \kappa y_0)\kappa u_0^2 \left(\frac{1 - \kappa y_0}{1 - \kappa y}\right)^3 \simeq (1 - \kappa y_0)^4 \kappa u_0^2 (1 + 3\kappa y)$$

$$\simeq (1 - \kappa y_0)^4 \kappa u_0^2 + 3\kappa^2 u_0^2 y.$$

(4) The equation of motion (5.12) can therefore be rewritten as:

$$\frac{d^2 y}{dt^2} + (\omega^2 + 3\kappa^2 u_0^2)y + \kappa u_0^2 (1 - \kappa y_0)^4 = 0. \tag{5.13}$$

Let us introduce the notations $\Omega^2 = \omega^2 + 3\kappa^2 u_0^2$ and $\bar{y} = -\kappa u_0^2 (1 - \kappa y_0)^4/\Omega^2$. Equation (5.13) is nothing but the equation of motion of an harmonic oscillator centered at $y = \bar{y}$, with frequency $\Omega > \omega$:

$$\frac{d^2(y - \bar{y})}{dt^2} + \Omega^2(y - \bar{y}) = 0.$$

The general solution of this equation is:

$$y = \bar{y} + (y_0 - \bar{y})\cos\Omega t + \frac{\dot{y}_0}{\Omega}\sin\Omega t.$$

We deduce two effects caused by the guide curvature: first, the effective angular frequency for the transverse motion is increased because of the centrifugal force contribution, and, second, this oscillation is not centered on the guide axis any more, but is displaced by a distance $\bar{y}$ further out from the guide axis. The larger the incident velocity, the further away the center of the transverse oscillation, as one might expect. Curved guides for cold neutral atoms have been investigated experimentally in recent years[4].

---

[4]See e.g. J.A. Sauer, M.D. Barrett, and M.S. Chapman, *Storage ring for neutral atoms*, Phys. Rev. Lett. **87**, 270401 (2001).

5.4   Role of non-linearities ***

We consider the one-dimensional motion (along x) of a particle of mass
m around its *stable* equilibrium position at $x = 0$. We denote by $U(x)$
the potential energy of the particle. We choose, without loss of generality,
$U(0) = 0$.

Harmonic approximation

What is the value of $U'(0)$?   What is the sign of $U''(0)$?   (we note
$U'(x) = \mathrm{d}U/\mathrm{d}x$, $U''(x) = \mathrm{d}^2U/\mathrm{d}x^2 \ldots$).   Show that the motion around
the equilibrium position is harmonic when the oscillations are of small am-
plitude.

Beyond the harmonic approximation

In this section, we study the corrections to the harmonic approximation,
first qualitatively, and then quantitatively. We restrict ourselves to the
terms of order three and four in the expansion of the potential $U(x)$.

(1) Write down the expression of the potential energy up to fourth order.
    In the case where $U(x)$ is an even function, show that the third-order
    term necessarily vanishes.
(2) Separating the cases of the third and fourth order, discuss how the
    average position of the particle and the oscillation period are modified
    by the non-linearities.
(3) Show that if we have only the third-order term, the equation of motion
    reads:

$$\ddot{x} + \omega_0^2 x + \alpha x^2 = 0. \qquad (5.14)$$

In order to determine, to first order in α, the role of this non-linearity,
we look for a solution of Eq. (5.14) of the form:

$$x(t) = a\cos(\omega t) + \delta x(t) \qquad \text{with} \qquad \delta x(t) = \delta_0 + \delta x_0 \cos(2\omega t).$$

Comment.   Why is there a term oscillating at 2ω in this solution?
Calculate ω, δ_0 and δx_0 as a function of α, a and ω_0.

(4) With only a fourth-order term, show that the equation of motion reads:

$$\ddot{x} + \omega_0^2 x + \beta x^3 = 0. \qquad (5.15)$$

To determine the role of the non-linear term to first order in β we look for a solution of Eq. (5.15) of the form:

$$x(t) = a\cos(\omega t) + \delta x(t) \qquad \text{with} \qquad \delta x(t) = \delta x_0 \cos(3\omega t). \qquad (5.16)$$

Comment on the form of this solution. Why is there a term oscillating at 3ω appearing? Calculate ω and δx_0 as a function of β, a and ω_0.

Non-linear mixing

In this section, we study the time evolution of the size of an atom cloud evolving in a potential well for which we take into account the non-harmonic corrections to fourth order: we have $U'''(0) = 0$, but $U^{(4)}(0) \neq 0$. The atoms have no initial velocity, and their initial positions are $x(0) = a$, where a has a Gaussian probability distribution:

$$\Pi(a) = \frac{1}{\sqrt{2\pi}\Delta}e^{-a^2/2\Delta^2}.$$

(1) Calculate, for a single atom, $x^2(a, t)$.
(2) Calculate for the atomic ensemble $\langle x^2\rangle(t)$, where

$$\langle x^2\rangle(t) = \int_{-\infty}^{\infty} \Pi(a)x^2(a, t)\,da,$$

first in the absence of non-linearities, and then taking them into account. Comment.

Mathematical reminder:

$$\cos^2 x = \frac{1}{2}(1 + \cos 2x)$$

$$\cos^3 x = \frac{1}{4}(3\cos x + \cos 3x)$$

$$\cos a \cos b = \frac{1}{2}\cos(a - b) + \frac{1}{2}\cos(a + b)$$

$$\int_{-\infty}^{\infty} a^{2p}e^{-\lambda a^2}\,da = \frac{1}{\lambda^{p+1/2}}\,\Gamma\left(p + \frac{1}{2}\right) \qquad \text{if} \qquad \text{Re}(\lambda) > 0.$$

The Euler Γ function takes on the following remarkable values:

$$\Gamma\left(\frac{1}{2}\right) = \sqrt{\pi}, \qquad \Gamma\left(\frac{3}{2}\right) = \frac{\sqrt{\pi}}{2},$$

$$\Gamma\left(\frac{5}{2}\right) = \frac{3\sqrt{\pi}}{4}, \qquad \Gamma\left(\frac{7}{2}\right) = \frac{15\sqrt{\pi}}{8}.$$

Solution

Harmonic approximation

By definition of an equilibrium position, $U'(0) = 0$ (no force is acting on the particle at $x = 0$). For a *stable* equilibrium, $U''(0) > 0$. We expand $U(x)$ around the equilibrium position:

$$U(x) = U(0) + xU'(0) + \frac{x^2}{2}U''(0) + \ldots = \frac{x^2}{2}U''(0) + \ldots$$

Newton's second law reads:

$$m\frac{\mathrm{d}^2 x}{\mathrm{d}t^2} = -\frac{\mathrm{d}U}{\mathrm{d}x} \simeq -xU''(0). \tag{5.17}$$

The equation of motion can be written as $\ddot{x} + \omega_0^2 x = 0$. We thus have harmonic oscillations around the equilibrium position $x = 0$, with the angular frequency ω_0 defined by $\omega_0^2 = U''(0)/m$. For $U''(0) < 0$, the solutions of Eq. (5.17) are diverging exponentials, revealing the unstable character of the solution.

Beyond the harmonic approximation

(1) The expansion of $U(x)$ up to fourth order reads:

$$U(x) = \frac{x^2}{2}U''(0) + \frac{x^3}{6}U'''(0) + \frac{x^4}{24}U^{(4)}(0).$$

If $U(x)$ is even, only even powers of x appear in the expansion, whence $U'''(0) = 0$.

(2) The average position is always $x = 0$ for an even potential. Therefore, if one includes for instance the fourth-order term, the average position is not changed by the non-linearity [Fig. 5.9(b)]. On the contrary, the third-order term breaks the symmetry between the $x > 0$ and $x < 0$ regions, and, depending on the sign of $U'''(0)$, the particle spends more time on one or the other side. The average position is shifted towards the weakest part of the trap, thus towards $x < 0$ if $U'''(0) > 0$ [Fig. 5.9(a)]. In the latter configuration, the particle spends more time on the $x < 0$ side, but less on the $x > 0$ side, than in the absence of non-linearities. We therefore do not expect, to first order, a change in the oscillation period. With a fourth-order non-linearity, preserving the parity of $U(x)$, the oscillation period is shorter if $U^{(4)}(0) > 0$ since the potential well is tighter in this case.

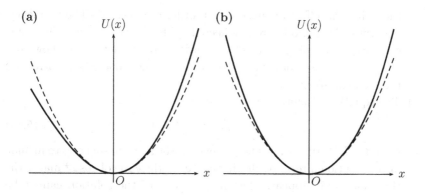

Fig. 5.9 (a) *Potential well $U(x)$ with a cubic term. Here $U'''(0) > 0$, the potential well is weaker in the region $x < 0$ than in the region $x > 0$. The particle thus spends more time in the region $x < 0$. The average position is thus negative.* (b) *Potential $U(x)$ with a quartic correction. In this case $U^{(4)}(0) > 0$, the potential well is tighter than in the harmonic approximation. The oscillation period is thus shorter. The dashed line is the parabola corresponding to the harmonic approximation.*

(3) The equation of motion reads:

$$\ddot{x} + \omega_0^2 x + \alpha x^2 = 0. \tag{5.18}$$

We have introduced $\alpha = U'''(0)/(2m)$. According to the previous question, the average position of the particle is nonzero, whence the constant term $\delta_0 \neq 0$. The presence of the quadratic term x^2 suggests to include in the solution a term oscillating at 2ω. Indeed, squaring a term $\cos \omega t$ makes the second harmonic appear (see the mathematical reminder). The frequency ω of the solution must equal ω_0, at least to first order in α. It thus makes sense to look for a correction of the form: $\delta x(t) = \delta_0 + \delta x_0 \cos(2\omega t)$. By injecting the suggested solution into Eq. (5.18), we get:

$$\delta \ddot{x} + \omega_0^2 \delta x + [\omega_0^2 - \omega^2] a \cos \omega t + \frac{\alpha a^2}{2}(1 + \cos 2\omega t) = 0.$$

Substituting the expression of δx and equating to zero the coefficients of 1, $\cos(\omega t)$, and $\cos(2\omega t)$, we obtain:

$$\omega = \omega_0, \qquad \delta_0 = -\frac{\alpha a^2}{2\omega_0^2}, \qquad \text{et} \qquad \delta x_0 = \frac{\alpha a^2}{6\omega_0^2}.$$

We can check that the minus sign in the expression of δ_0 makes sense. Indeed, as $\delta_0 = -U'''(0)a^2/(4m\omega_0^2)$, if $U'''(0) > 0$ the average position of the particle is shifted towards $x < 0$, i.e. towards the weakest side of the well. This effect appears at first order as expected. The frequency, and thus the period, are unaffected to first order in α. The same reasoning, but going up to second order, would allow us to show that the frequency is modified to second order (i.e. one has a correction proportional to α^2).

(4) The equation of motion now reads:

$$\ddot{x} + \omega_0^2 x + \beta x^3 = 0, \tag{5.19}$$

with $\beta = U^{(4)}(0)/(6m)$. The presence of the x^3 term suggests to include in the solution a term oscillating at 3ω, as the cube of $\cos \omega t$ makes the third harmonic appear. The frequency ω of the solution cannot be equal to ω_0, even to first order in β, as was discussed qualitatively above. The potential well remains symmetrical, and thus there cannot be a constant term, unlike in the previous case (such a term would imply that the particle spends different times in the regions $x < 0$ and $x > 0$, which is impossible for an even potential). It is thus natural to look for a correction of the form: $\delta x(t) = \delta x_0 \cos(3\omega t)$. Injecting the solution into Eq. (5.19), we have:

$$\delta \ddot{x} + \omega_0^2 \delta x + \left[\omega_0^2 - \omega^2 + \frac{3\beta a^2}{4} \right] a \cos \omega t + \frac{\beta a^3}{4} \cos 3\omega t = 0.$$

We then deduce, equating to zero the coefficients of $\cos(\omega t)$ et $\cos(3\omega t)$:

$$\omega^2 = \omega_0^2 + \frac{3\beta a^2}{4}, \quad \text{or} \quad \omega = \omega_0 \left(1 + \frac{3\beta a^2}{8\omega_0^2} \right), \quad \text{and} \quad \delta x_0 = \frac{\beta a^3}{32\omega_0^2}.$$

The change in frequency implies a modification of the period:

$$T = \frac{2\pi}{\omega} = \frac{2\pi}{\omega_0} \left(1 - \frac{3\beta a^2}{8\omega_0^2} \right).$$

It is interesting to recover this result by a purely energetic approach. For that, we obtain a constant of motion by multiplying Eq. (5.19) by $\dot{x}$ and integrating over time. Noting a the turning point of the trajectory (point where $\dot{x} = 0$), we get:

$$\dot{x}^2 = \omega_0^2 (a^2 - x^2) \left[1 + \frac{\beta}{2\omega_0^2} (a^2 + x^2) \right].$$

We thus deduce the expression of the oscillation period T:

$$\frac{T}{4} = \int_0^a \frac{dx}{\sqrt{\omega_0^2 (a^2 - x^2) \left[1 + \beta(a^2 + x^2)/(2\omega_0^2) \right]}}.$$

By making an expansion to first order in β under the integral, we obtain:

$$\frac{T}{4} = \frac{1}{\omega_0} \int_0^a \frac{dx}{\sqrt{a^2 - x^2}} - \frac{\beta}{4\omega_0^3} \int_0^a \frac{a^2 + x^2}{\sqrt{a^2 - x^2}}\, dx.$$

We can calculate the last two integrals using the change of variables $x = a \sin \psi$, with $\psi \in [0, \pi/2]$:

$$\int_0^a \frac{dx}{\sqrt{a^2 - x^2}} = \int_0^{\pi/2} d\psi = \frac{\pi}{2}$$

and

$$\int_0^a \frac{a^2 + x^2}{\sqrt{a^2 - x^2}}\, dx = a^2 \int_0^{\pi/2} (1 + \sin^2 \psi)\, d\psi = \frac{3\pi}{4} a^2.$$

Therefore we get

$$T = T_0 \left(1 - \frac{3\beta a^2}{8\omega_0^2} \right).$$

We recover the expected behavior: if $U^{(4)}(0) > 0$, i.e. $\beta > 0$, the potential well including the non-linearity is more confining than in the harmonic approximation, and the oscillation period is therefore shorter. Moreover, one loses the isochronism of the oscillations. Indeed, because of the a^2 term, the oscillation period depends on the energy of the particle. Finally, this calculation can be applied to take into account the first correction in the differential equation of a simple pendulum:

$$\ddot{\theta} + \omega_0^2 \sin \theta \simeq \ddot{\theta} + \omega_0^2 \left(\theta - \frac{\theta^3}{6} \right) = 0.$$

Here we have $\beta = -\omega_0^2/6$, and thus

$$T = T_0 \left(1 + \frac{\theta_0^2}{16} \right).$$

This result, also known as *Borda's equation*, is also obtained in Problem 5.1, where it is compared to experimental results.

Non-linear mixing

(1) The goal of this question is to account quantitatively for the smearing out of the oscillations in the radius of a trapped atomic cloud under the influence of non-linearities. This results from the fact that the oscillation period is different for all particles, since it depends explicitly on the amplitude a.

From the previous results, $x(t) = a\cos\omega t + \epsilon a^3 \cos 3\omega t$, where $\epsilon = \beta/(32\omega_0^2)$ and ω also depends on a: $\omega(a) = \omega_0 + 3\beta a^2/(8\omega_0)$. Using the mathematical reminder, one easily obtains the following expression for $x^2(t)$:

$$x^2(a,t) = \frac{a^2 + \epsilon^2 a^6}{2} + \frac{a^2}{2}\cos 2\omega t$$
$$+ \frac{\epsilon^2 a^6}{2}\cos(6\omega t) + \epsilon a^4 \cos 2\omega t + \epsilon a^4 \cos 4\omega t.$$

(2) In the absence of non-linearity ($\epsilon = 0$), $x^2(a,t)$ oscillates at $2\omega_0$, and so does the average $\langle x^2\rangle(t)$ since ω_0 does not depend on a. This collective oscillation mode is known as the *monopole mode*, or *breathing mode*. The calculation of $\langle x^2\rangle$ is more complicated when one takes into account the non-linearity which changes the oscillation frequency. We need to calculate integrals of the form:

$$I_{n,p} = \int_{-\infty}^{\infty} a^{2p} \cos[2n\omega(a)t]e^{-a^2/2\Delta^2}\,da.$$

Let us first calculate an integral $J_{n,p}$ such as $I_{n,p} = \mathrm{Re}(J_{n,p})$. Using the mathematical reminder:

$$J_{n,p} \equiv e^{2in\omega_0 t} \int_{-\infty}^{\infty} a^{2p}e^{-\lambda_n a^2}\,da = \frac{e^{2in\omega_0 t}}{\lambda_n^{p+1/2}}\Gamma\left(p + \frac{1}{2}\right),$$

with

$$\lambda_n = \frac{1}{2\Delta^2} - \frac{3in\beta t}{4\omega_0} = \rho_n e^{-i\theta_n}.$$

We finally deduce the expression of $I_{n,p}$:

$$I_{n,p}(t) = \frac{\Gamma(p + 1/2)}{\rho_n^{p+1/2}}\cos\left[2n\omega_0 t + \left(p + \frac{1}{2}\right)\theta_n(t)\right].$$

The solution can be expressed in terms of the $I_{n,p}(t)$:

$$\langle x^2\rangle(t) = \langle x^2\rangle^{(0)}(t) + \epsilon\langle x^2\rangle^{(1)}(t) + \epsilon^2\langle x^2\rangle^{(2)}(t) + \ldots$$

with

$$\langle x^2\rangle^{(0)}(t) = \frac{\Delta^2}{2} + \frac{1}{2\sqrt{2\pi}\Delta}I_{1,1}(t),$$

$$\langle x^2\rangle^{(1)}(t) = \frac{1}{\sqrt{2\pi}\Delta}\left[I_{2,2}(t) + I_{2,1}(t)\right],$$

$$\langle x^2\rangle^{(2)}(t) = \frac{15\Delta^6}{2} + \frac{1}{2\sqrt{2\pi}\Delta}I_{3,3}(t).$$

We have expanded $\langle x^2 \rangle(t)$ in powers of ϵ. The main contribution is of zeroth-order[5] in ϵ:

$$\langle x^2 \rangle^{(0)}(t) = \frac{\Delta^2}{2} + \frac{\Delta^2}{2\left(1 + \dfrac{9\beta^2\Delta^4 t^2}{4\omega_0^2}\right)^{3/4}} \cos\left(2\omega_0 t + \frac{3}{2}\theta_1(t)\right).$$

We recover a fingerprint of the oscillation at twice the trap frequency, characteristic of the breathing mode, with the term in $2\omega_0 t$ of the argument of the cosine. The zeroth-order solution $\langle x^2 \rangle^{(0)}(t)$ shows all the features of *nonlinear mixing*: the non-exponential decay of the oscillation amplitude, and the presence of a *phase slip* term $\theta_1(t)$. The decrease in the oscillation amplitude is the signature of the smearing out of the oscillations due to the non-isochronism of individual oscillations. The terms appearing in the corrections proportional to ϵ and ϵ^2 are of the same nature, but their amplitude decays faster than that of the zeroth-order term.

---

[5]Of zeroth order in ϵ, but not in β, as ω also depends on β.

Chapter 6

A Few Theorems in Mechanics

6.1 The virial theorem **

Lemma: Show that the time average, over a long enough time, of the time derivative of a *bounded* function f, is equal to zero.

Let us consider a particle of mass m subjected to a conservative central force $\boldsymbol{F}(\boldsymbol{r})$ with the corresponding potential

$$U(r) = Ar^\lambda, \tag{6.1}$$

where A and λ are constants. We assume that the motion is limited to a bounded region in space. We denote by $\langle X \rangle$ the time average of a quantity X.

(1) Show, using the result of the lemma above, that the average of the kinetic energy fulfills:

$$\langle E_{\mathrm{k}} \rangle = -\frac{1}{2} \langle \boldsymbol{r} \cdot \boldsymbol{F} \rangle . \tag{6.2}$$

(2) Deduce from Eq. (6.2) the following relation between the averages of the kinetic and potential energies:

$$2 \langle E_{\mathrm{k}} \rangle = \lambda \langle U \rangle . \tag{6.3}$$

This is called the virial theorem[1].

(3) Link $\langle E_{\mathrm{k}} \rangle$ and $\langle U \rangle$ to the total energy E of the particle.

(4) For the specific cases $\lambda = 2$ and $\lambda = -1$, show that one recovers well-known results. Interpret also the case $\lambda \to \infty$.

---

[1] One calls *virial*, from the latin *vis*, force, the right-hand side of Eq. (6.2).

Solution

Lemma: By definition, the time average of f' over a time T reads:

$$\langle f' \rangle = \frac{1}{T} \int_0^T f'(t)\,\mathrm{d}t = \frac{f(T) - f(0)}{T} ,$$

which indeed goes to 0 when $T \to \infty$ since we have assumed that f is bounded.

(1) The average value of the kinetic energy reads:

$$\langle E_k \rangle = \frac{1}{2}m \langle v^2 \rangle = \frac{1}{2}m \left\langle v \cdot \frac{\mathrm{d}r}{\mathrm{d}t} \right\rangle .$$

Using the formula giving the derivative of a product, we get

$$\langle E_k \rangle = \frac{1}{2}m \left\langle \frac{\mathrm{d}}{\mathrm{d}t}(v \cdot r) - r \cdot \frac{\mathrm{d}v}{\mathrm{d}t} \right\rangle .$$

The first term vanishes according to the lemma (indeed $v{\cdot}r$ is a bounded function, since, on the one hand, the motion is restricted to a finite region of space, and, on the other hand, the velocity cannot reach infinity). We thus have:

$$\langle E_k \rangle = -\frac{1}{2} \left\langle r \cdot m\frac{\mathrm{d}v}{\mathrm{d}t} \right\rangle = -\frac{1}{2} \langle r \cdot F \rangle ,$$

where we have used Newton's second law in order to write the last equality.

(2) The potential U corresponds to the force field:

$$F = -\boldsymbol{\nabla} U = -A\lambda r^{\lambda-2} r ,$$

thus we have $\langle F \cdot r \rangle = -A\lambda \langle r^\lambda \rangle = -\lambda \langle U \rangle$. Finally, we get:

$$2 \langle E_k \rangle = \lambda \langle U \rangle ,$$

i.e. the virial theorem.

(3) We have a conservative force, and therefore the total mechanical energy E is a conserved quantity:

$$E = E_k + U \qquad \Longrightarrow \qquad E = \langle E_k \rangle + \langle U \rangle .$$

Using the virial theorem, this yields:

$$\begin{cases} \langle E_k \rangle = \dfrac{\lambda}{\lambda+2} E , \\[2mm] \langle U \rangle = \dfrac{2}{\lambda+2} E . \end{cases} \tag{6.4}$$

(4) The case $\lambda = 2$ corresponds to a harmonic oscillator. The equations of motion read

$$\ddot{\boldsymbol{r}} + \omega^2 \boldsymbol{r} = 0$$

with $\omega^2 = 2A/m$. The trajectory thus has the equation:

$$\boldsymbol{r}(t) = \boldsymbol{r}_0 \cos(\omega t) + \frac{\boldsymbol{v}_0}{\omega} \sin(\omega t).$$

The calculation of $\langle U \rangle = \langle m\omega^2 r^2/2 \rangle$ and $\langle E_{\mathrm{k}} \rangle = \langle m\dot{r}^2/2 \rangle$ is immediate if one notices that[2] $\langle \cos^2 \rangle = \langle \sin^2 \rangle = 1/2$ and $\langle \cos\sin \rangle = 0$. We do find in the end that $\langle U \rangle = \langle E_{\mathrm{k}} \rangle = m(v_0^2 + \omega^2 r_0^2)/4$, in agreement with the virial theorem for $\lambda = 2$.

The case $\lambda = -1$ corresponds to the Kepler problem. To have a bounded motion, the interaction must be attractive: $U(r) = -k/r$ with $k > 0$. Let us recall a few results whose proof can be found in any textbook on classical mechanics: the motion occurs within a plane (one has a central force, thus the angular momentum is constant, its magnitude being $L = mr^2\dot{\theta}$), the trajectory is an ellipse of polar equation (the origin being located at the focus, i.e. at the attractive center)

$$r(\theta) = \frac{p}{1 + e\cos\theta},$$

where $p = L^2/mk$ and e, the *eccentricity*. Substituting the equation of the trajectory into the total energy E (one has $E < 0$ for this bound state), one gets the following important relationship:

$$e^2 - 1 = \frac{2L^2}{mk^2} E. \tag{6.5}$$

Let us calculate the time average of U. The motion is periodic with

---

[2] Let us show that the average of $\cos^2$ over one period is $1/2$. By definition,

$$\langle \cos^2 \rangle = \frac{1}{T} \int_0^T \cos^2 \omega t \, dt = \frac{1}{2\pi} \int_0^{2\pi} \frac{1 + \cos(2x)}{2} \, dx,$$

where we have made the change of variable $x = \omega t$, with $\omega = 2\pi/T$, and where we have used the trigonometric identity $\cos^2 x = [1 + \cos(2x)]/2$. Performing the integration is then elementary and gives $1/2$. The proof that $\langle \sin^2 \rangle = 1/2$ is similar (it amounts to changing the origin of time). Finally, the proof that $\langle \sin\cos \rangle = 0$ is obvious by noticing that $\sin x \cos x = \sin(2x)/2$.

period T; we thus have:

$$\langle U \rangle = \frac{1}{T} \int_0^T \frac{-k}{r(t)} \, dt$$

$$= \frac{-km}{TL} \int_0^{2\pi} r(\theta) \, d\theta \qquad \text{(for } dt = mr^2 d\theta/L)$$

$$= \frac{-kmp}{TL} \int_0^{2\pi} \frac{1}{1 + e\cos\theta} \, d\theta$$

$$= \frac{-kmp}{TL} \frac{2\pi}{\sqrt{1 - e^2}} \,, \qquad\qquad (6.6)$$

where the last integral is calculated by the change of variable $\xi = \tan(\theta/2)$. The period T fulfills Kepler's third law $T^2/a^3 = 4\pi^2 m/k$, where $a = p/(1 - e^2)$ is the semi-major axis of the ellipse. One then deduces $T = 2\pi L^3/[mk^2(1 - e^2)^{3/2}]$. Comparing with Eq. (6.6), this yields:

$$\langle U \rangle = \frac{mk^2}{L^2}(e^2 - 1) = 2E \,,$$

where the last equality was obtained using Eq. (6.5). We have therefore proven $\langle U \rangle = 2E$, which is indeed the second equation of (6.4) for $\lambda = -1$.

Finally, the case $\lambda \to \infty$ corresponds to the motion in an infinitely deep square potential well (writing $U(r)$ as $U_0(r/r_0)^\lambda$, it is clear that in the limit $\lambda \to \infty$ we have $U(r) = 0$ if $r < r_0$ and $U(r) = \infty$ otherwise). The interpretation of Eq. (6.4) is then very clear, as in this case the potential energy always vanishes, and the total energy is equal to the kinetic energy.

♦ **Remark.** The virial theorem can be proven for a class of potentials more general than Eq. (6.1), namely those that are *homogeneous of degree* λ, i.e. that fulfill:

$$\forall \alpha \in \mathbb{R}, \quad U(\alpha x, \alpha y, \alpha z) = \alpha^\lambda U(x, y, z).$$

This allows us to generalize the result, Eq. (6.3), for instance to the case of an *anisotropic* harmonic oscillator (with, in this case, $\lambda = 2$). For a proof, see § 10 of L. Landau and E. Lifchitz, *Mechanics*, Butterworth-Heinemann (1982). Note that the virial theorem can be extended to systems made of N particles, both in classical and in quantum mechanics[3].

---

[3] A recent example in the field of ultracold atoms can be found in F. Werner, *Virial theorems for trapped cold atoms*, Phys. Rev. A **78**, 025601 (2008), also available online at http://arxiv.org/abs/0803.3277.

6.2 Adiabatic invariants ***

In this exercise, one first shows, in the case of two simple examples, that a mechanical oscillator characterized by a parameter λ which is varied slowly (with respect to the oscillation period), possesses *adiabatic invariants*, i.e. that a certain combination $I(E, \lambda)$ of the energy E and of λ stays constant, although E and λ vary in time. The third part is devoted to the study of the general case.

One-dimensional harmonic oscillator

We consider a one-dimensional harmonic oscillator, with potential energy

$$V(x) = \frac{1}{2}m\omega^2 x^2 .$$

One varies the angular frequency ω, very slowly with respect to the oscillation period.

(1) Write down the total energy E of the particle as a function of the position x, the velocity v, and the frequency ω.
(2) What is the differential $\mathrm{d}E$ of the energy E?
(3) Using Newton's second law, show that

$$\mathrm{d}E = m\omega^2 x^2 \frac{\mathrm{d}\omega}{\omega} .$$

(4) By averaging the previous expression over an oscillation period, show that $I(E, \omega) = E/\omega$ is an adiabatic invariant of the problem.
(5) Interpret this result in the framework of quantum mechanics.
(6) What is the nature of the trajectory in phase space (x, v)? Show that the invariance of E/ω implies that the area enclosed by the trajectory in phase space is also invariant.

Infinite potential well

We now consider a square potential well of infinite depth and width ℓ:

$$V(x) = 0 \quad (0 \leqslant x \leqslant \ell)$$
$$V(x) = +\infty \quad (x < 0 \text{ or } x > \ell)$$

whose width $\ell(t)$ is varied slowly, by moving one of the "walls" (see Fig. 6.1).

(1) By writing down the variation of the energy E of the particle upon reflection on the moving wall, exhibit an adiabatic invariant $I(E, \ell)$.

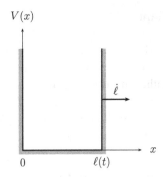

Fig. 6.1   *Infinite potential well whose width ℓ varies.*

(2) Interpret the result in the framework of quantum mechanics.
(3) What is the nature of the trajectory in phase space? Show that the invariance of I is equivalent to that of the area enclosed by the trajectory in phase space over one period.

General case

We will show that for a one-dimensional oscillator with a potential $V(x, \lambda)$ whose parameter λ varies adiabatically, the adiabatic invariant reads:

$$I = \frac{1}{2\pi} \oint p(x)\,\mathrm{d}x \tag{6.7}$$

where $p(x)$ is the momentum of the particle, the integral being taken over a full period of the motion.

(1) Show that with this formula one does recover the results obtained above for the harmonic oscillator and for the infinite square well.
(2) Show that the invariance of I corresponds to the invariance of the area enclosed in phase space (x, p) by the closed trajectory corresponding to one period.
(3) By analogy with the proof for the case of the harmonic oscillator, show the invariance of Eq. (6.7).

Solution

One-dimensional harmonic oscillator

(1) The total mechanical energy of the particle obviously reads:
$$E = \frac{1}{2}mv^2 + \frac{1}{2}m\omega^2 x^2 .$$

(2) Its differential dE is obtained immediately:
$$dE = mv\,dv + m\omega^2 x\,dx + m\omega x^2\,d\omega .$$

(3) Newton's second law $m\ddot{x} = F = -m\omega^2 x$ also reads $dv = -\omega^2 x dt$. Since $dx = vdt$, the first two terms in dE cancel each other. We finally end up with:
$$dE = m\omega^2 x^2 \frac{d\omega}{\omega} .$$

(4) Averaging the previous expression over an oscillation period, we get the variation δE of the energy on a period:
$$\delta E = \left\langle m\omega^2 x^2 \right\rangle \frac{\delta\omega}{\omega} .$$

The factor $\delta\omega/\omega$ was taken out of the average, as it varies very little on the timescale of one period (this is the very definition of the assumption of adiabaticity). Moreover, it is not difficult to see that the average value of $m\omega^2 x^2$ over one period is equal to the energy E of the particle in the absence of a perturbation[4]. We thus have
$$\frac{\delta E}{E} = \frac{\delta\omega}{\omega} ,$$
or, upon integration,
$$\frac{E}{\omega} = \text{const.}$$

(5) In quantum mechanics, one shows that the energy of particle in a harmonic potential reads $E = \hbar\omega(n + 1/2)$, where $n \in \mathbb{N}$ is the quantum number characterizing the quantum state of the particle. If ω varies adiabatically, we know from the adiabatic theorem that the particle remains in the same quantum state: n stays constant, and consequently $E/\omega = \text{const.}$

---

[4]For the harmonic motion, we have $x(t) = A\cos(\omega t + \varphi)$, and thus $m\omega^2 \langle x^2(t)\rangle = m\omega^2 A^2/2$ (the average of $\cos^2$ over one period is $1/2$, see footnote 2 on page 83). In the same way, $m\langle \dot{x}^2(t)\rangle = m\omega^2 A^2/2$. The average kinetic energy and potential energy are thus equal to half the total energy. This is a particular case of the *virial theorem* (see Problem 6.1).

(6) We have $E = mv^2/2 + m\omega^2 x^2/2$. This defines in the plane (x, v) an ellipse of semi-axes $a = \sqrt{2E/m}/\omega$ (in x) and $b = \sqrt{2E/m}$ (in v), as shown in Fig. 6.2(a). The "area" of this ellipse is

$$A = \pi ab = 2\pi \frac{E}{m\omega}.$$

Thus, when ω varies adiabatically, the ellipse gets distorted but keeps a constant area.

Infinite potential well

(1) The particle of energy E arrives on the moving wall with a velocity $v_0 = \sqrt{2E/m}$. The wall velocity is $\dot{\ell}$, and the assumption of an adiabatic change (i.e. the relative variation $\delta\ell/\ell$ of the well width ℓ over a period $T = 2\ell/v_0$ of oscillation is small) implies that $\dot{\ell} \ll v_0$. In the wall frame, the particle is reflected elastically: its velocity simply changes sign, going from $v_0 - \dot{\ell}$ to $-v_0 + \dot{\ell}$. In the lab frame, the velocity of the particle thus becomes $-v_0 + 2\dot{\ell}$. The new energy of the particle is therefore:

$$E + \delta E = \frac{1}{2}m(-v_0 + 2\dot{\ell})^2 \simeq E - 2mv_0\dot{\ell},$$

where the last equality is valid to first order in $\dot{\ell}/v_0$. Writing

$$\dot{\ell} = \frac{\delta\ell}{T} = \frac{\delta\ell}{2\ell/v_0},$$

we finally have:

$$\delta E = -mv_0^2 \frac{\delta\ell}{\ell} = -2E\frac{\delta\ell}{\ell}.$$

Integrating, we get:

$$I(E, \ell) = E\ell^2 = \text{const.}$$

(2) In quantum mechanics, the energy levels of a particle of mass m in an infinite potential well of width ℓ are given by:

$$E = \frac{\hbar^2\pi^2 n^2}{2m\ell^2},$$

with $n \in \mathbb{N}^*$. If ℓ varies adiabatically, the particle remains in the same quantum state, therefore n stays constant, and $E\ell^2 = \text{const.}$

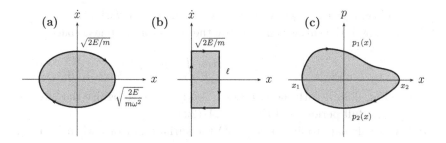

Fig. 6.2 *Trajectories in phase space for a harmonic oscillator* (a) *and for the infinite potential well* (b). *In* (c), *illustration of Eq.* (6.8).

(3) In phase space, the point (x, v) moves, in between two collisions ($0 \leqslant x \leqslant \ell$), at a speed $v = \pm\sqrt{2E/m} = \text{const.}$ During collisions, the representative point of the particle in phase space jumps immediately from (x, v) to $(x, -v)$. The trajectory in phase space is thus a rectangle with sides ℓ (in x) and $2\sqrt{2E/m}$ (in v), as shown on Fig. 6.2(b). The area of this rectangle is thus $\propto \ell\sqrt{E} = \sqrt{I}$, and remains constant as ℓ varies adiabatically.

General case

(1) The total energy reads $E = p^2/(2m) + V(x, \lambda)$; one deduces that $p(x) = \pm\sqrt{2m}\,(E - V(x, \lambda))^{1/2}$. For the case of the harmonic oscillator, one has:

$$I = \frac{1}{2\pi}4\sqrt{2m}\int_0^{x_{\mathrm{m}}}\sqrt{\frac{1}{2}m\omega^2(x_{\mathrm{m}}^2 - x^2)}\,\mathrm{d}x\,,$$

with $E = m\omega^2 x_{\mathrm{m}}^2/2$. We used the symmetries of the trajectory in phase space (x, p) to reduce the calculation of the integral over the full period to an integral over only a quarter of a period. Introducing $u = x/x_{\mathrm{m}}$, we find:

$$I = \frac{2}{\pi}m\omega x_{\mathrm{m}}^2\int_0^1\sqrt{1 - u^2}\,\mathrm{d}u = \frac{E}{\omega}\,,$$

since the remaining integral is equal to $\pi/4$ (surface of a quarter of a disk of radius 1). One thus recovers the adiabatic invariant E/ω. For the infinite well, $p(x) = \pm\sqrt{2mE}$, and thus

$$I = \frac{1}{2\pi}2\int_0^{\ell}\sqrt{2mE}\,\mathrm{d}x = \frac{\ell\sqrt{2mE}}{\pi}\,.$$

We obtain, as expected, that $E\ell^2$ is an adiabatic invariant.

(2) The area of the surface enclosed by the trajectory in phase space reads

$$\mathcal{A} = \int_{x_1}^{x_2} p_1(x)\, dx - \int_{x_1}^{x_2} p_2(x)\, dx, \tag{6.8}$$

[see Fig. 6.2(c)] with, here, $p_2(x) = -p_1(x)$. We recognize the integral over the full period, and thus $I = \mathcal{A}/(2\pi)$.

(3) Let us calculate to first order in $\delta\lambda$ the variation δI of I when λ varies:

$$2\pi\,\delta I = \delta \oint p(x)\, dx = \oint \sqrt{2m}\,\delta \left[(E - V(x,\lambda))^{1/2} \right] dx.$$

Performing the variation, we obtain:

$$2\pi\,\delta I = \oint \sqrt{2m}\, \frac{\delta E - \dfrac{\partial V}{\partial \lambda}\delta\lambda}{2\left(E - V(x,\lambda)\right)^{1/2}}\, dx. \tag{6.9}$$

Now the energy of the particle is $E = mv^2/2 + V(x,\lambda)$. Its differential therefore reads:

$$dE = mv\, dv + \frac{\partial V}{\partial x}\, dx + \frac{\partial V}{\partial \lambda}\, d\lambda. \tag{6.10}$$

Newton's second law

$$m\frac{dv}{dt} = -\frac{\partial V}{\partial x}$$

implies that the first two terms of Eq. (6.10) cancel out. The energy variation thus reads

$$\delta E = \frac{\partial V}{\partial \lambda}\, \delta\lambda.$$

Substituting into Eq. (6.9), we deduce that, to first order in $\delta\lambda$, $\delta I = 0$. The integral I, called the *action integral*, is thus an adiabatic invariant. In fact, one can show that if the variation of $\lambda(t)$ is smooth enough (of class C^∞), then the adiabatic invariant is conserved with an accuracy which increases exponentially when $\dot\lambda$ decreases. The concept of adiabatic invariant played a major role during the early days of quantum mechanics at the beginning of the twentieth century (Bohr and Sommerfeld's "old quantum theory" postulated the quantization of the action integral; this quantization rule can be derived from Schrödinger equation using the WKB semi-classical limit of quantum mechanics — see the reminder below).

Reminder: WKB quantization rule

Within the framework of the semi-classical approximation, also called the WKB (Wentzel–Kramers–Brillouin) approximation, the energy levels in a smooth one-dimensional potential well are obtained using the quantization rule

$$\oint p(x)\,dx = \left(n + \frac{1}{2}\right)h,$$

where h is Planck's constant, the momentum $p(x) = \sqrt{2m\left[E - V(x)\right]}$ is calculated for the energy E, and n is a non-negative integer. The integral is taken over a full period of the motion (see e.g. A. Messiah, *Quantum mechanics*, North Holland, New York (1961), chapter 6.2.). This equation defines the energy levels E_n, and gives the same expression as the solutions of the Schrödinger equation for the case $n \gg 1$. One recognizes in the l.h.s. the action integral discussed above. The reader is invited to calculate by the WKB method the energy levels of the harmonic oscillator, and to compare with the exact results.

6.3 Lax form and constants of motion ***

In this exercise, we develop a systematic method to obtain the constants of motion for a problem of classical mechanics. The method is illustrated in one dimension.

The harmonic oscillator

In the first part, we consider a one-dimensional harmonic oscillator. The corresponding potential reads:

$$U(x) = \frac{1}{2}m\omega^2 x^2.$$

(1) Write down the corresponding equation of motion. Show that this second-order linear differential equation can be rewritten as a set of two coupled first-order linear differential equations.

(2) We introduce the 2×2 matrix:

$$\mathsf{L} = \sqrt{m}\begin{pmatrix} v & \omega x \\ \omega x & -v \end{pmatrix}.$$

Show that the set of first-order linear differential equations derived in question (1) can be put in the following matrix form:

$$\frac{d\mathsf{L}}{dt} = [\mathsf{L}, \mathsf{M}], \tag{6.11}$$

where the matrix M has constant coefficients. Give the expression for these coefficients. The notation [L, M] refers to the *commutator* of the two matrices L and M, defined by [L, M] = LM − ML.

(3) Show that the matrix relation (6.11) implies that the quantities Tr(L^n), where n is a positive integer, are constants of motion.

The Calogero–Sutherland model

This one-dimensional model describes the dynamics of N particles of mass m that interact through the long-range two-body potential $U(r) = mg^2/r^2$, where r is the relative distance between the interacting particles, and g a coupling constant that characterizes the strength of the repulsive interactions. The N-body interaction energy therefore reads:

$$U(x_1,\ldots,x_N) = \frac{1}{2} \sum_{i \neq j} \frac{mg^2}{(x_i - x_j)^2}.$$

(1) Determine the set of equations of motion for each particle.
(2) We introduce the $N \times N$ matrix L = Δ + Λ, defined by

$$\Delta_{kj} = \delta_{kj} v_j, \quad \Lambda_{kj} = \begin{cases} 0, & k = j \\ i\dfrac{g}{(x_k - x_j)} & k \neq j \end{cases}$$

where δ_{kj} is the Kronecker symbol[5], and i the complex number $i^2 = -1$. Show that there exists a matrix M that allows one to rewrite the set of equations of motion in the form:

$$\frac{\mathrm{d}L}{\mathrm{d}t} = [L, M]. \tag{6.12}$$

Determine the explicit expression of the coefficients of this matrix.

(3) Deduce from the previous question that the quantities $C_n =$ Tr(L^n)$/n!$ are constants of motion. Give the explicit expression of C_1 and C_2.
(4) How many independent constants of motion are there in the Calogero–Sutherland model?
(5) To solve the equations of motion, we introduce the auxiliary matrix X whose coefficients are $X_{kj} = x_k \delta_{kj}$. Show that

$$\frac{\mathrm{d}X}{\mathrm{d}t} = [X, M] + L. \tag{6.13}$$

---

[5]Defined by $\delta_{jj} = 1$, and $\delta_{jk} = 0$ if $j \neq k$.

(6) We introduce the scalar quantities:

$$E_n(t) = \frac{1}{n!}\mathrm{Tr}(\mathsf{X}\mathsf{L}^{n-1}).$$

Calculate the explicit expression of $E_n(t)$ as a function of time t and of the coefficients C_n.

(7) Apply the previous formalism to the case $N = 3$. Comment.

Solution

The harmonic oscillator

(1) The equation of motion is readily obtained from Newton's second law:

$$m\frac{d^2x}{dt^2} = -\frac{\partial U}{\partial x} = -m\omega^2 x.$$

This equation can be recast as the following set of two coupled first-order linear differential equations:

$$\frac{dv}{dt} = -\omega^2 x \quad \text{and} \quad \frac{dx}{dt} = v. \tag{6.14}$$

(2) Let us consider the most general form for the matrix M:

$$\mathsf{M} = \begin{pmatrix} a & b \\ c & d \end{pmatrix}.$$

The commutator reads:

$$[\mathsf{L}, \mathsf{M}] = \sqrt{m}\begin{pmatrix} (c-b)\omega x & 2bv + (d-a)\omega x \\ -2cv + (a-d)\omega x & (b-c)\omega x \end{pmatrix}. \tag{6.15}$$

Furthermore, using Eq. (6.14), we obtain the expression for $d\mathsf{L}/dt$:

$$\frac{d\mathsf{L}}{dt} = \sqrt{m}\omega\begin{pmatrix} -\omega x & v \\ v & \omega x \end{pmatrix}. \tag{6.16}$$

We deduce from Eqs (6.15) and (6.16) a possible choice for the coefficients of M ($a = d = 0$ and $b = -c = \omega/2$) to obtain the relation (6.11) of the text. From the equation of motion one thus gets the matrix relation (6.11). Conversely, the matrices L and M combined with the relation (6.11) allow one to derive the equation of motion. These two formulations are therefore equivalent.

(3) Let us first calculate dL^n/dt:

$$\frac{dL^n}{dt} = \sum_{k=0}^{n} L^k \frac{dL}{dt} L^{n-1-k} = \sum_{k=0}^{n} L^k [L, M] L^{n-1-k}$$

$$= \sum_{k=0}^{n} (L^{k+1} M L^{n-1-k} - L^k M L^{n-k}).$$

Indeed, dL/dt does not commute with L. Let us recall at this stage the well-known property of the trace $\mathrm{Tr}(AB) = \mathrm{Tr}(BA)$, valid for any pair of square matrices A and B. We thus have:

$$\frac{d\mathrm{Tr}(L^n)}{dt} = \sum_{k=0}^{n} \mathrm{Tr}(L^{k+1} M L^{n-1-k} - L^k M L^{n-k})$$

$$= \sum_{k=0}^{n} [\mathrm{Tr}(ML^n) - \mathrm{Tr}(ML^n)] = 0.$$

We have thus shown that $\mathrm{Tr}(L^n)$ is a conserved quantity. This property is not specific to the peculiar example considered in the first part of the exercise. It originates from the rewriting of Newton's second law in the matrix form (6.11). For the harmonic oscillator, one has $L^2 = (2E)\mathrm{Id}$, where Id is the identity matrix and E the total mechanical energy of the particle, so that:

$$\mathrm{Tr}(L^{2n+1}) = 0, \quad \text{and} \quad \mathrm{Tr}(L^{2n}) = 2(2E)^n.$$

We recover a well-known result: the energy is a conserved quantity. However, as we shall see in the second part of this exercise, this method allows one to find non-trivial constants of motion for more complex situations.

The Calogero–Sutherland model

(1) The equations of motion read:

$$\begin{cases} \dfrac{dv_k}{dt} = -\dfrac{\partial U}{\partial x_k} = 2 \displaystyle\sum_{l \neq k} \dfrac{g^2}{(x_k - x_l)^3}, \\[2ex] \dfrac{dx_k}{dt} = v_k. \end{cases}$$

(2) We are looking for a matrix M whose coefficients fulfill:

$$\frac{dL_{kj}}{dt} = \sum_{p=1}^{N} (L_{kp} M_{pj} - M_{kp} L_{pj}). \tag{6.17}$$

Let us first calculate the coefficients dL_{kj}/dt using the equations of motion:

$$\begin{cases} \dfrac{dL_{kk}}{dt} = \dfrac{d\Delta_{kk}}{dt} = \dfrac{dv_k}{dt} = 2\sum_{l \neq k} \dfrac{g^2}{(x_k - x_l)^3}, \\[3mm] \dfrac{dL_{kj}}{dt} = \dfrac{d\Lambda_{kj}}{dt} = -i\dfrac{g}{(x_k - x_j)^2}(v_k - v_j), \quad k \neq j. \end{cases} \qquad (6.18)$$

The last relation for non-diagonal coefficients can be rewritten in the form:

$$\frac{dL_{kj}}{dt} = L_{kk}M_{kj} - L_{jj}M_{jk} \qquad (6.19)$$

where

$$M_{kj} = -i\frac{g}{(x_k - x_j)^2} \quad \text{with } k \neq j,$$

since $L_{kk} = v_k$. Let us choose the values of the diagonal coefficients M_{jj} so that we obtain a relation of the type of Eq. (6.17). It will then remain to check that this formulation also gives the expected results for the diagonal elements of dL/dt. Let us write explicitly the first part of the r.h.s. of Eq. (6.17):

$$\sum_{p=1}^{N} L_{kp}M_{pj} = \sum_{p=1}^{N} \Delta_{kp}M_{pj} + \sum_{p=1}^{N} \Lambda_{kp}M_{pj}$$

$$= \Delta_{kk}M_{kj} + \Lambda_{kj}M_{jj} + \sum_{p \neq (j,k)}^{N} \Lambda_{kp}M_{pj}.$$

We deduce the expression for the elements (k, j) of the commutator $[L, M]$:

$$[L, M]_{kj} = (L_{kk}M_{kj} - L_{jj}M_{kj}) + \Lambda_{kj}(M_{jj} - M_{kk})$$

$$+ \sum_{p \neq (j,k)}^{N} (\Lambda_{kp}M_{pj} - M_{kp}\Lambda_{pj}). \qquad (6.20)$$

Equations (6.19) and (6.20) are consistent with the matrix form of (6.12) for the equations of motion if the diagonal elements of the matrix

M fulfills the equations:

$$M_{jj} - M_{kk} = -\frac{1}{\Lambda_{kj}} \sum_{p \neq (j,k)}^{N} (\Lambda_{kp} M_{pj} - M_{kp} \Lambda_{pj})$$

$$= ig \sum_{p \neq (j,k)}^{N} \frac{(x_k - x_j)(x_k - x_j - 2x_p)}{(x_k - x_p)^2 (x_p - x_j)^2}$$

$$= ig \sum_{p \neq (j,k)}^{N} \frac{(x_k - x_p)^2 - (x_p - x_j)^2}{(x_k - x_p)^2 (x_p - x_j)^2}.$$

We therefore define M_{jj} by:

$$M_{jj} = ig \sum_{p \neq j} \frac{1}{(x_p - x_j)^2}.$$

It remains to check that this choice is consistent with the equations for the time evolution of the diagonal elements dL_{kk}/dt :

$$\frac{dL_{kk}}{dt} = \sum_{p=1}^{N} (L_{kp} M_{pk} - M_{kp} L_{pk})$$

$$= \sum_{p \neq k} (L_{kp} M_{pk} - M_{kp} L_{pk})$$

$$= \sum_{p \neq k} (\Lambda_{kp} M_{pk} - M_{kp} \Lambda_{pk})$$

$$= \sum_{p \neq k} \left(\frac{ig}{(x_k - x_p)} \frac{(-ig)}{(x_k - x_p)^2} - \frac{(-ig)}{(x_k - x_p)^2} \frac{ig}{(x_p - x_k)} \right)$$

$$= 2 \sum_{p \neq k} \frac{g^2}{(x_k - x_p)^3}.$$

In this way, we recover the set of equations (6.18). In summary, we have two equivalent formulations for the equations of motion.

(3) As in the first part, the matrix formulation (6.11) of the equations of motion immediately yields an ensemble of C_n constants of motion. For C_1 and C_2, we find:

$$C_1 = \sum_{i=1}^{N} v_i,$$

$$C_2 = \frac{1}{2} \sum_i (L^2)_{ii} = \frac{1}{2} \sum_{i,j} L_{ij} L_{ji} = \frac{1}{2} \sum_i \Delta_i^2 + \frac{1}{2} \sum_{i \neq j} |\Lambda_{ij}|^2$$

$$= \sum_{i=1}^{N} \frac{1}{2} v_i^2 + \frac{1}{2} \sum_{i \neq j} \frac{g^2}{(x_i - x_j)^2} = \frac{E}{m}.$$

C_1 gives the conservation of the momentum of the center of mass. The fact that this quantity is a constant can be shown directly from the equations of motion by summing all the individual accelerations dv_i/dt. The forces that result from two-body interaction are internal forces of the ensemble of N particles, and therefore do not affect the center of mass motion. The constant C_2 gives the conservation of energy.

(4) Consider the case $N = 1$. The constant of motion is the energy, which links the particle velocity to its position. For a one-dimensional system made of simply one particle, there is at most one conserved quantity. Indeed, if there were two independent conserved quantities, the position x and the velocity v would be constant in time. This reasoning can be readily generalized for larger numbers of particles: a one-dimensional system made of N particles has at most N independent conserved quantities[6]. The property

$$\frac{d\mathrm{Tr}(\mathsf{L}^n)}{dt} = 0$$

means that all symmetric polynomials in the eigenvalues of L are conserved, and therefore that the eigenvalues of L are conserved quantities. The constants of motion are thus given here by the C_n $(1 \leq n \leq N)$ that are independent polynomials of degree n in x and v. The C_n with $n > N$ can be expressed as polynomials in the C_n $(1 \leq n \leq N)$.

(5) The matrix expression (6.13) suggests to calculate the components of the commutator $[\mathsf{X}, \mathsf{M}]$. For the off-diagonal elements, we find:

$$
\begin{aligned}
[\mathsf{X}, \mathsf{M}]_{kj} &= \sum_i (\mathsf{X}_{ki}\mathsf{M}_{ij} - \mathsf{M}_{ki}\mathsf{X}_{ij}) \\
&= \mathsf{X}_{kk}\mathsf{M}_{kj} - \mathsf{M}_{kj}\mathsf{X}_{jj}) \\
&= \mathsf{M}_{kj}(x_k - x_j) \\
&= -\mathsf{L}_{kj} \text{ if } k \neq j.
\end{aligned}
$$

For the diagonal elements, one has $[\mathsf{X}, \mathsf{M}]_{kk} = 0$, so that $d\mathsf{X}_{kk}/dt = dx_k/dt = v_k = \mathsf{L}_{kk}$. We have thus shown the relation (6.13).

---

[6]The maximum number of conserved quantities N_m^c of system made of N particles depends on the dimensionality of the space where the particles evolve. In a three-dimensional space $(d = 3)$, $N_m^c = 3N$. For our case, $d = 1$ and $N_m^c = N$.

(6) Let us calculate the time derivative:

$$n! \frac{dE_n}{dt} = \text{Tr}\left(\frac{dX}{dt}L^{n-1} + X\frac{dL^{n-1}}{dt}\right)$$

$$= \text{Tr}([X, M]L^{n-1} + L^n + X[L^{n-1}, M])$$

$$= \text{Tr}(L^n) + \text{Tr}(XML^{n-1} - MXL^{n-1} + XL^{n-1}M - XML^{n-1})$$

$$= \text{Tr}(L^n) + \text{Tr}(XL^{n-1}M - MXL^{n-1}) = \text{Tr}(L^n).$$

We obtain the expression for the quantities $E_n(t)$:

$$\frac{dE_n}{dt} = C_n \qquad \Longrightarrow \qquad E_n(t) = E_n(0) + C_n t.$$

The quantities $E_n(0)$ contain the information on the initial conditions.

(7) The case $N = 3$ is the simplest one for which the theory that we have developed gives a non-trivial constant of motion C_3, i.e. that is different from the conservation of the center of mass momentum or the total energy. We find:

$$C_1 = v_1 + v_2 + v_3,$$

$$C_2 = \frac{1}{2}\left(v_1^2 + v_2^2 + v_3^2\right)$$
$$+ g^2\left(\frac{1}{(x_1 - x_3)^2} + \frac{1}{(x_3 - x_2)^2} + \frac{1}{(x_1 - x_2)^2}\right),$$

$$C_3 = \frac{1}{6}\left(v_1^3 + v_2^3 + v_3^3\right)$$
$$+ \frac{g^2}{2}\left(\frac{v_1 + v_3}{(x_1 - x_3)^2} + \frac{v_3 + v_2}{(x_3 - x_2)^2} + \frac{v_1 + v_2}{(x_1 - x_2)^2}\right).$$

We obtain also the $E_n(t)$:

$$E_1(t) = x_1 + x_2 + x_3,$$

$$E_2(t) = \frac{1}{2}(v_1 x_1 + v_2 x_2 + v_3 x_3),$$

$$E_3(t) = \frac{1}{6}\left(v_1^2 x_1 + v_2^2 x_2 + v_3^2 x_3\right)$$
$$+ \frac{g^2}{6}\left(\frac{x_1 + x_2}{(x_1 - x_2)^2} + \frac{2x_1}{(x_1 - x_3)^2}\right.$$
$$\left. + \frac{2x_2}{(x_2 - x_3)^2} + \frac{1}{x_3 - x_1} + \frac{1}{x_3 - x_2}\right).$$

We have in total six equations, clearly independent, that link the variables $x_1(t)$, $x_2(t)$, $x_3(t)$, $v_1(t)$, $v_2(t)$ and $v_3(t)$. Three of them result from the constants of motion C_1, C_2 et C_2, and the three others from the $E_i(t)$:

$$E_1(t) = E_1(0) + C_1 t,$$
$$E_2(t) = E_2(0) + C_2 t,$$
$$E_3(t) = E_3(0) + C_3 t.$$

One finally gets a set of six non-linear equations that enables one to extract the position and the velocity of each particle at any time.

To conclude, the method studied here permits one to solve exactly the Calogero–Sutherland model at any time by solving an algebraic set of equations, without computing the full time evolution to get the positions and the velocities of the N interacting particles at a given time. This exercise illustrates the efficiency of algebraic approaches in classical mechanics.

Chapter 7

Collisions

7.1 Collisions and thermalization *

The question we would like to answer with this exercise is the following: can a one-dimensional gas, made of impenetrable hard spheres trapped in a potential well, thermalize?

Consider two identical particles in a one-dimensional harmonic potential:

$$U(x) = \frac{1}{2}m\omega^2 x^2.$$

The two particles are initially at rest, one in the region $x < 0$, and the other in the region $x > 0$. Their initial positions are arbitrary (see Fig. 7.1).

(1) Determine the abscissa of the first elastic collision between the two particles.

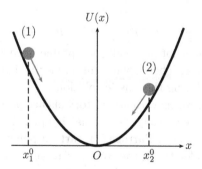

Fig. 7.1 *The two particles, initially at rest, evolve in a one-dimensional harmonic potential well.*

101

(2) Represent in phase space the elastic collision of the two particles. Deduce the successive positions where the collisions occur. Conclude.

(3) Explain the origin of thermalization for systems with a larger dimensionality.

Solution

(1) The particles have initially no velocity. One of them starts at an arbitrary position in the region $x < 0$, and the other one at an arbitrary position in the region $x > 0$. As they evolve in a harmonic potential, they reach the bottom O of the trap at the same time $t = T/4$, where $T = 2\pi/\omega$ is the oscillation period, because of the isochronism of harmonic oscillations. The first collision therefore occurs at $x = 0$.

(2) To determine the velocity of each particle after the collision, we use the conservation of momentum and kinetic energy during the collision process:

$$p_1 + p_2 = p_1' + p_2' \qquad \text{and} \qquad p_1^2 + p_2^2 = p_1'^2 + p_2'^2,$$

where p_1 and p_2 are the momenta of particle one and two, respectively, before the collision, and are supposed to be known. The quantities with a prime refer to the momenta after the collision. The previous set of equations can be recast in the form

$$p_1 p_2 = p_1' p_2' \qquad \text{and} \qquad p_1 + p_2 = p_1' + p_2'.$$

As (p_1, p_2) and (p_1', p_2') have the same product and sum, they are the solutions of the same quadratic equation. There are thus two possible solutions:

$$p_1 = p_1' \quad \text{and} \quad p_2 = p_2' \qquad \text{or} \qquad p_2 = p_1' \quad \text{and} \quad p_1 = p_2'.$$

The first solution is not physically acceptable since we consider impenetrable hard spheres. The second solution means that particles exchange their momenta during the elastic collision.

In Fig. 7.2(a), we show the trajectory of each particle in phase space. We have deliberately exaggerated what happens in the collision region for pedagogical reasons. Particle (2) starts at point A at the abscissa x_2^0 with no initial velocity. In the course of its motion, the absolute value of its velocity increases (the velocity is negative), and the particle goes in the direction of the bottom of the trap, following the curve that goes from A to B in the phase portrait. Going from A to B requires

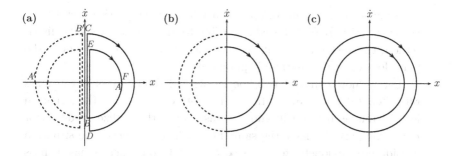

Fig. 7.2 (a) *Trajectories in phase space* $(x, \dot{x})$ *for particles (1) (dashed line) and (2) (solid line). The region of the collision is deliberately magnified to show the transfer of momentum that occurs during the collision.* (b) *Same curve, but without the details on the collision region.* (c) *Same curve as in* (b)*, but for indistinguishable particles; there is no more distinction between dashed and solid lines.*

a time $T/4$. During the same amount of time, particle (1) follows the curve that connects the points A' and B'. The collision occurs, and the particles exchange their velocities. Particle (2) thus reaches the point C after the collision, and follows the path from C to D in a time $T/2$. At D, particle (2) collides again with particle (1), and evolves from E to F, and so on. During the time evolution, particle (1) always remains in the region $x < 0$, while particle (2) remains in the region $x > 0$. Figure 7.2(b) summarizes the trajectories on the phase portrait for each particle. The collisions always occur at the bottom of the trap (this is due to the fact that initially the particles had no velocity) at times τ_n:

$$\tau_n = \frac{T}{4} + n\frac{T}{2}, \qquad n \in \mathbb{N}.$$

(3) In one dimension, particles simply exchange their momenta during the collision. The phase portrait for indistinguishable particles is therefore that of two independent particles that do not "see" each other, as illustrated in Fig. 7.2(c) where the same solid line is used for both particles.

   In this exercise, we chose a harmonic potential for the sake of simplicity, but the result, that originates from local conservation laws for the collision process, holds for any confining potential. The collision acts as an exchange of the role played by the two particles, there is therefore no possible thermalization in a one-dimensional gas. Such a behavior has been observed recently with ultracold atoms trapped in one dimension by means of optical lattices[1]. If the particles have their velocities initially distributed according to an arbitrary velocity distribution, this distribution will remain unchanged after one period if the particles evolve in a harmonic potential.

Let us now consider two species with different masses, the mass ratio being denoted by r. If $r = 1$, $r \to \infty$, or $r \to 0$, the situation is the same as previously and the gas cannot thermalize. Indeed, for $r = 1$ the particles have the same mass which is exactly the situation considered previously; for $r \to \infty$ or $r \to 0$, one species has a mass much larger than the other one, and is therefore essentially insensitive to the presence of the light species. One may wonder what happens for an intermediate value of the ratio r. Actually, a thermalization-like process occurs in this situation and there exists a value $r^\star$ of the mass ratio for which the time needed for thermalization is minimum[2].

What happens if the dimensionality of space is larger? Let us consider an elastic collision that occurs in a space of dimension d. We know the velocities of the two colliding particles, and want to determine their values after the collision. Each velocity vector has d components, so that we have to determine $2d$ components. The conservation laws yield $d + 1$ independent equations: one is provided by the conservation of kinetic energy, and the other ones come from the conservation of the d components of the total momentum of the two particles. We conclude that we have in general, for $d > 1$, more quantities to determine than equations. This can be called the degree of indetermination of the collision, defined by $2d - (d+1) = d-1$. For $d = 1$, the degree of indetermination is zero since we have the same number of unknown quantities as conservation law equations. We have indeed found a unique solution in this case. However, for $d > 1$, there are not enough equations to determine a unique solution for the velocities of the particles after the collision.

In two dimensions, the conservation laws ensure the conservation of the norm of the relative velocity vector $\|\boldsymbol{v}_2' - \boldsymbol{v}_1'\| = \|\boldsymbol{v}_2 - \boldsymbol{v}_1\|$. The angle between these two vectors can take any value, and, in the absence of further information about the collision process, is therefore assumed to be uniformly distributed on the interval $[0, 2\pi]$.

---

[1]T. Kinoshita, T.R. Wenger, and D.S. Weiss, *A quantum Newton's cradle*, Nature **440**, 900 (2006).

[2]P. Mohazzabi and J.R. Schmidt, *Relaxation of one-dimensional binary mixtures*, Phys. Rev. E **55**, 6881 (1997).

Similarly, in three dimensions, one has to introduce two arbitrary spherical angles[3]. These undetermined quantities in the treatment of elastic collisions allow for the exchange of energy, and the collision is not a mere exchange of the role played by each particle as in the one-dimensional case. This possibility of exchanging energy in dimension $d > 1$ thus enables the thermalization of the gas.

7.2 Minimum distance of approach *

We consider two identical charged particles of charge q and mass m, that are constrained to move on the axis (Ox). The first particle comes from infinity, and has an initial velocity v_0; the other particle is initially at rest. Calculate the minimum distance of approach between the two particles. Comment.

Evaluate numerically the minimum distance of approach of two colliding protons, the moving one having an incident kinetic energy of 1 MeV (we recall the values of the proton mass: $m_p = 1.67 \times 10^{-27}$ kg, of the elementary charge: $q = 1.6 \times 10^{-19}$ C, and of the vacuum permittivity: $\varepsilon_0 = 8.85 \times 10^{-12}$ F $\cdot$ m^{-1}).

Solution

There are different ways to solve this problem. Let us use for instance Newton's second law for both particles:

$$m\ddot{x}_1 = \frac{q^2}{4\pi\varepsilon_0(x_1 - x_2)^2} \quad \text{and} \quad m\ddot{x}_2 = -\frac{q^2}{4\pi\varepsilon_0(x_1 - x_2)^2},$$

where x_1 refers to the position of particle (1), and x_2 of particle (2). Let us introduce the relative distance $u = x_1 - x_2$. From the previous equations, we deduce

$$m\ddot{u} = \frac{2q^2}{4\pi\varepsilon_0 u^2}.$$

Energy conservation for this equation of motion is obtained by multiplying it by $\dot{u}$, and by integrating over time:

$$m\ddot{u}\dot{u} = \frac{2q^2}{4\pi\varepsilon_0 u^2}\dot{u} \quad \Longleftrightarrow \quad \frac{d}{dt}\left(\frac{1}{2}m\dot{u}^2\right) = -\frac{d}{dt}\left(\frac{q^2}{2\pi\varepsilon_0 u}\right).$$

---

[3]In quantum mechanics, these undetermined angles are replaced by angular probability distributions whose exact form depends on the interaction potential and on the collision energy.

We thus find:

$$\frac{1}{2}m\dot{u}^2 + \frac{q^2}{2\pi\varepsilon_0 u} = E_0.$$

The initial conditions yield the value of the constant $E_0 = mv_0^2/2$. The minimum distance of approach $u_{\min}$ is obtained from the turning point of the trajectory associated with the relative motion, i.e. at $\dot{u} = 0$: $u_{\min} = q^2/(2\pi\varepsilon_0 E_0)$. The larger the incident energy, the lower the minimum distance of approach. This is the reason why particle accelerators communicate a very large energy to the particles before they collide, in order to probe their interactions at short distances.

Numerically, we find:

$$u_{\min} = \frac{q}{E_0}\frac{q}{2\pi\varepsilon_0} = \frac{1}{10^6}\frac{1.6\times 10^{-19}}{2\pi\,8.85\times 10^{-12}} \simeq 2.9\times 10^{-15} \text{ m}.$$

An incident energy on the orders of a few MeV thus permits one to study the strong nuclear force, that has a very short range, on the order of 10^{-15} m, i.e. the nucleus size. From this respect, particle accelerators can be considered to be "microscopes".

♦ **Remark.** The use of non-relativistic mechanics requires that $v_0 \ll c$, i.e. $E_0 \ll m_{\mathrm{p}}c^2$. This is indeed the case here, since we have $E_0 = 1$ MeV and $m_{\mathrm{p}}c^2 \simeq 938$ MeV.

▶ **Further reading.** The Large Hadron Collider (LHC) that started to operate at CERN near Geneva in 2008 is the most powerful particle accelerator ever built, and is designed to accelerate protons with an energy of 7 TeV against each other. This lies obviously in the ultra-relativistic domain. A very interesting and accessible review of the physics and technology of the LHC is available on the internet at www.nature.com/nature/supplements/insights/hadroncollider/.

7.3 Landau's criterion of superfluidity ⋆⋆

Below the critical temperature $T_{\mathrm{c}} \simeq 2.2$ K, liquid helium flows without any viscosity: one says that it becomes *superfluid*. A "test" particle can move in the liquid without undergoing any drag force, provided its velocity stays below a critical velocity v_{c}. These fascinating properties were discovered in the thirties, some twenty years after the achievement of the liquefaction of helium. The goal of this problem is to study a microscopic interpretation of these properties that was put forward in the forties by the Soviet physicist Lev Landau.

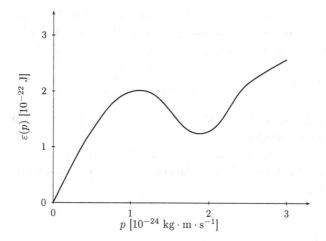

Fig. 7.3 *Spectrum $\varepsilon(p)$ of the elementary excitations of superfluid helium.*

Let us consider a particle of mass M moving in the fluid with a velocity $\boldsymbol{v}$. The particle will experience a drag force only if it can transfer part of its energy to the fluid. We assume that this interaction with the fluid consists, at the microscopic level, in creating an "elementary excitation", which can be described as a particle[4] having a momentum $\boldsymbol{p}$ and an energy $\varepsilon(p)$.

(1) Write down the conservation of energy and of momentum for the process where the "test" particle creates an elementary excitation (we will denote by $\boldsymbol{v}'$ the final velocity of the test particle).

(2) Show that energy and momentum conservation imply

$$v > \min_{p} \frac{\varepsilon(p)}{p}.$$

Deduce the value of the critical velocity below which the liquid is superfluid.

(3) For an ideal gas, the elementary excitations are simply free particles: one thus has $\varepsilon(p) = p^2/(2m)$, where m is the mass of the particles of the gas. Is an ideal gas superfluid?

---

[4]These elementary excitations are in fact called *quasi-particles* because they have properties similar to those of "real" particles, but also some specific properties: for instance, their number is not conserved. They correspond to a collective motion of the particles of the medium. *Phonons* in a crystalline lattice, which are simply collective oscillations (normal modes) of the atoms of the lattice, are an example of quasi-particles.

(4) In the framework of quantum mechanics, one can show that for a gas of weakly interacting bosons of mass m, the dispersion relation of the elementary excitations is given by the so-called *Bogoliubov* spectrum:

$$\varepsilon(p) = p\sqrt{c^2 + \frac{p^2}{4m^2}}.$$

Comment on this form. What does the constant c represent physically? Is this gas superfluid?

(5) Figure 7.3 shows the spectrum $\varepsilon(p)$ of the elementary excitations of superfluid helium. What is the critical velocity according to Landau's criterion?

Solution

(1) Energy conservation gives:

$$\frac{Mv^2}{2} = \frac{Mv'^2}{2} + \varepsilon(p),$$

and by momentum conservation we have

$$M\boldsymbol{v} = M\boldsymbol{v'} + \boldsymbol{p}.$$

(2) Eliminating $\boldsymbol{v'}$ between the two above equations yields

$$\frac{Mv^2}{2} - \varepsilon(p) = \frac{M}{2}\left(\boldsymbol{v} - \frac{\boldsymbol{p}}{M}\right)^2,$$

and thus we must have, for this process:

$$\boldsymbol{v}\cdot\boldsymbol{p} = \varepsilon(p) + \frac{p^2}{2M} \geq \varepsilon(p). \tag{7.1}$$

Let us assume that $v < v_c$, where we define the critical velocity as

$$v_c \equiv \min_{p} \frac{\varepsilon(p)}{p}.$$

Then $\boldsymbol{v}\cdot\boldsymbol{p} = vp\cos\theta < v_c p < \varepsilon(p)$, and Eq. (7.1) cannot be fulfilled. No quasi-particle can thus be created and the gas is superfluid.

(3) For the ideal gas, we have $\varepsilon(p) = p^2/(2m)$. Therefore $\varepsilon(p)/p = p/(2m)$, whose minimum is equal to zero (reached for $p = 0$). We deduce that for an ideal gas $v_c = 0$, and thus an ideal gas cannot be superfluid.

(4) The Bogoliubov spectrum

$$\varepsilon(p) = p\sqrt{c^2 + \frac{p^2}{4m^2}}$$

reduces, for small momenta $p \ll mc$, to the linear relation $\varepsilon = pc$. This dispersion relation characterizes *phonons*, i.e. the excitation quanta of a sound wave. To see this, we just need to write $\varepsilon = \hbar\omega$ and $p = \hbar k$, which gives $\omega = ck$, the dispersion relation of a (non-dispersive) wave, where c is the propagation velocity (here the speed of sound). For momenta large compared to mc, we get $\varepsilon \simeq p^2/(2m) + mc^2$, i.e. the dispersion relation of a free particle $p^2/(2m)$ (within a constant energy offset mc^2). Let us calculate the critical velocity:

$$v_c = \min_p \frac{\varepsilon(p)}{p} = \min_p \sqrt{c^2 + \frac{p^2}{4m^2}} = c.$$

According to Landau's criterion, an interacting Bose gas is therefore a superfluid[5], the speed of sound being the critical velocity.

(5) In order to determine graphically the critical velocity given by Landau's criterion, we must find the minimal slope of the straight line going through the origin and through the point $M(p, \varepsilon(p))$ of the dispersion relation (as its slope is $\varepsilon(p)/p$). This line is shown as a dashed line on Fig. 7.4. We get $v_c \simeq 60$ m $\cdot$ s^{-1}. In fact, the experimentally measured value is much smaller, on the order of a few cm $\cdot$ s^{-1} at most, when the motion involves macroscopic objects. The above value for the critical velocity is, however, obtained experimentally when the test particle is microscopic, such as an ion (see for instance D.R. Allum, P.V.E. McClintock, and A. Phillips, *The breakdown of superfluidity in liquid* ^4He: *an experimental test of Landau's theory*, Phil. Trans. R. Soc. Lond. A **284**, 179 (1977)).

▶ **Further reading.** For an introduction to the properties of superfluid helium, one can read chapter 13 of K. Huang's *Statistical mechanics*, Wiley, New York (1963). The original paper by L.D. Landau, J. Phys. USSR **5**, 71 (1941), is reproduced in *Collected papers of L.D. Landau*, edited by D. ter Haar, Pergamon, London (1965).

---

[5]This conclusion has been checked experimentally, see e.g. A.P. Chikkatur, A. Görlitz, D.M. Stamper-Kurn, S. Inouye, S. Gupta, and W. Ketterle, *Suppression and enhancement of impurity scattering in a Bose-Einstein condensate*, Phys. Rev. Lett. **85**, 483 (2000).

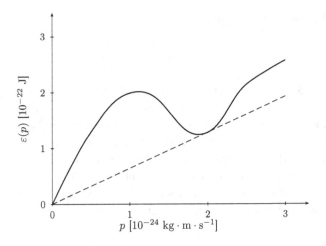

Fig. 7.4   *Spectrum $\varepsilon(p)$ of the elementary excitations of superfluid helium, together with the line of minimum slope giving Landau's critical velocity.*

7.4   Collision cross section **

Definition and properties

We consider the scattering of particles of mass m, arriving from infinity with a velocity v_∞, by a potential $U(\boldsymbol{r})$ localized in the neighborhood of the origin O. After the interaction, the direction of propagation of the particles is characterized by the angles (θ, φ) measured from the initial direction of propagation of the particles (see Fig. 7.5). Depending on the initial value of the transverse coordinates (b_x, b_y) of the particles (the so-called *impact parameters*), they will be scattered in different directions. We thus have a correspondance[6] $(b_x, b_y) \longrightarrow (\theta, \varphi)$.

We assume that the incident flux of particles (i.e. the number of incident particles per unit time and unit surface) is Φ. The number of particles scattered into the elementary solid angle $d\Omega$ per unit time, divided by the incident flux, is called the *differential scattering cross section* into this solid angle.

(1)   We denote the differential cross section by

$$\frac{d\sigma}{d\Omega}(\theta, \varphi).$$

What is the dimension of this quantity?

---

[6] The inverse correspondance might be multi-valued.

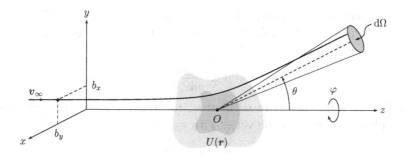

Fig. 7.5   *Scattering of a particle by a potential well $U(\boldsymbol{r})$.*

(2) We assume that the impact parameters are known as a function of the polar angles (θ, φ). Write the differential cross section $\mathrm{d}\sigma$ in terms of (b_x, b_y).

(3) We now assume that $U(\boldsymbol{r})$ is cylindrically symmetric around z. The impact parameter $b = \sqrt{b_x^2 + b_y^2}$ is thus sufficient to characterize the collision. What is then the expression of the differential cross section (which now depends only on θ)?

(4) The integral of the differential cross section over all directions of space is called the total cross section (and is noted σ). What is the value of σ for a spherically symmetric potential which vanishes outside a sphere of radius R? What is the value of σ for a potential which, on the contrary, extends to infinity?

Scattering by a hard sphere

We now consider the scattering by a hard sphere of radius R, i.e. a potential which is infinite for $r \leqslant R$, and vanishes when $r > R$. Calculate the differential cross section. Comment. What is the value of the total cross section?

The case of a power-law central force

We consider here the scattering by a repulsive central field with a power-law dependence on the distance from the center:

$$U(\boldsymbol{r}) = \frac{\alpha}{r^n}, \quad (\alpha > 0).$$

(1) Calculate the scattering angle θ as an integral, which one will not try to calculate in closed form, and show that θ depends only on the dimensionless quantity

$$\beta \equiv \frac{2\alpha}{mv_\infty^2 b^n}.$$

(2) Deduce from the above result that, for a given scattering angle, the differential cross section varies with v_∞ as:

$$\frac{d\sigma}{d\Omega} \propto v_\infty^{-4/n}.$$

(3) Calculate explicitly the differential cross section in the case of the Coulomb potential ($n = 1$). We recall the value of the following integral:

$$\int_0^a \frac{du}{\sqrt{(a - u)(u + 1/a)}} = 2\arctan a.$$

Scattering by a "soft" sphere

We consider the scattering by the following central potential:

$$U(\mathbf{r}) = \begin{cases} m\omega^2(R^2 - r^2)/2 & \text{if } r \leqslant R \\ 0 & \text{otherwise.} \end{cases}$$

(1) Calculate the scattering angle as a function of b.
(2) Which case, already studied above, do we recover in the limit of low-energy scattering $v_\infty \ll \omega R$?
(3) On the contrary, for the case of high-energy scattering $v_\infty \gg \omega R$, show that one can find the result very easily by a perturbative approach: the particle is only weakly deflected from its initial trajectory, and therefore one can calculate the transverse velocity acquired during the interaction with the potential by assuming that the particle still has a straight trajectory during that time.
(4) Calculate the maximal deflection angle in this high-energy limit.

Application to the atomic structure: the Thomson model and Rutherford's experiment

At the beginning of the 1900s, physicists knew that atoms contain electrons, and consequently also positive charges, so that they are neutral. They also knew quite an accurate value of the Avogadro number N_A, estimated by J. Loschmidt around 1865, which allowed them to calculate atomic sizes.

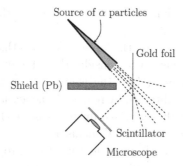

Source of α particles

Gold foil

Shield (Pb)

Scintillator

Microscope

Fig. 7.6 *Geiger and Marsden's experiment. The lead (Pb) shield prevents the alpha particles from reaching the scintillator directly from the source.*

In 1909, J.J. Thomson proposed the following model of the atom: a sphere of radius $R \sim 10^{-10}$ m, uniformly charged in volume, with a positive charge density, inside which the electrons are moving. Rutherford then suggested his students H. Geiger and E. Marsden perform the following experiment: a beam of alpha particles of energy $E_0 = 5.6$ MeV is directed towards a very thin gold foil, and a scintillating screen (observed using a microscope) allows for the detection of the alpha particles scattered by the foil at an angle $\theta \simeq \pi/2$ (see Fig. 7.6). The experiment shows that a significant number of particles is detected, showing unambiguously that the differential cross section for the scattering of an alpha particle by an atom of gold (atomic number $Z = 79$) is nonzero for $\theta = \pi/2$. We recall that alpha particles are helium nuclei (of charge $2e$, where $e \simeq 1.6 \times 10^{-19}$ C is the elementary charge, and of mass $4m$, where $m \simeq 1.66 \times 10^{-27}$ kg is the atomic mass unit).

(1) How can one estimate the size of an atom, knowing (like in 1900) N_A, the molar mass M of an element, and the density ρ of a crystal of the same element? What is its numerical value for gold ($M = 197$ g·mol^{-1} and $\rho = 19,300$ kg·m^{-3})?

(2) Using Gauss's theorem, give the expression of the electrostatic potential created by the charge distribution (disregarding the electrons).

(3) In order to study the scattering of alpha particles, why is it possible, in a first approximation, to neglect (i) the recoil of the gold nucleus, and (ii) the influence of the electrons?

(4) We assume here that the Thomson model is valid. Calculate the maximal deflection in the case where the alpha particle does not enter into

the charge distribution. Take $R = 10^{-10}$ m for the numerical calculation.

(5) In order to try and explain the large deflections that are experimentally observed, we are thus led to consider that the alpha particles that are strongly deflected enter into the charge distribution. Calculate the maximal deflection obtained by replacing the actual potential with that of a soft sphere as above; its parameters will be chosen such that both potentials coincide for $r < R$. Conclusion?

(6) Rutherford, in an article published in 1911, proposed to explain the large deflections by assuming that the positive charge of the atom is concentrated in a very tiny region — the nucleus. In the framework of this model, calculate, for the parameters of the Geiger and Marsden experiment, the minimal distance between the alpha particle and the nucleus. This gives an upper bound on the size of the atomic nucleus. How does it compare with the typical nuclear sizes known nowadays?

Solution

Definition and properties

(1) The cross section is the ratio of a number of particles per unit time (dimension T^{-1}) by a flux (dimension $L^{-2}T^{-1}$); its dimension L^2 is thus that of a surface (whence its name).

(2) Only the particles having impact parameters (b_x, b_y) will have a final velocity in the direction (θ, φ). Therefore, these particles having their impact parameters within $(b_x, b_x + db_x) \times (b_y, b_y + db_y)$ are the ones that will reach the solid angle $d\Omega$. Their number is $\Phi \, db_x \, db_y$ per unit time, and thus the cross section simply reads

$$\frac{d\sigma}{d\Omega} \, d\Omega = |db_x \, db_y| \,,$$

where we still need to express the differential elements db_x and db_x as a function of (θ, φ) (the absolute value is there to obtain a positive area even in the case where the differential elements, expressed as a function of the spherical angles, are negative). If the functions $b_x(\theta, \varphi)$ and $b_y(\theta, \varphi)$ are multi-valued, one needs to sum the above expression over all the determinations of $b_{x,y}$.

(3) For a potential which is invariant under rotation around z, we use spherical coordinates with polar axis z. Integrating over φ, we get

$|\mathrm{d}b_x \, \mathrm{d}b_y| = 2\pi b |\mathrm{d}b|$, and $\mathrm{d}\Omega = 2\pi \sin\theta \, \mathrm{d}\theta$. We thus have:

$$\frac{\mathrm{d}\sigma}{\mathrm{d}\Omega}(\theta) = \frac{b}{\sin\theta} \left| \frac{\mathrm{d}b}{\mathrm{d}\theta} \right|. \tag{7.2}$$

(4) Integrating $\mathrm{d}\sigma/\mathrm{d}\Omega$ over all the directions of space amounts to integrating over all the impact parameters giving a nonzero scattering angle; we find

$$\sigma = \iint \mathrm{d}b_x \, \mathrm{d}b_y = \begin{cases} \pi R^2 & \text{if } U(r) = 0 \text{ for } r > R, \\ \infty & \text{otherwise.} \end{cases}$$

♦ **Remarks.**

(i) One may wonder why one uses the differential cross section, expressed as a function of θ only, although measuring $\theta(b)$ (scattering angle versus impact parameter) is in principle sufficient to characterize the scattering properties of the potential. This is because experimentally, when one studies the scattering by microscopic scatterers, one has no control over b. With a uniform incident flux, one has a distribution of values of b (with a probability distribution $p(b) \, \mathrm{d}b \propto b \, \mathrm{d}b$), and one can *measure* only the distribution of the scattering angles θ. The differential cross section thus corresponds to what can naturally be measured in an experiment.

(ii) The concept of *total* cross section is not very meaningful in classical mechanics since, as we have just seen, it is either infinite, or equal to the geometrical section of the potential. However, in quantum mechanics, the total cross section is finite in general; this is why this concept is important. For instance, if we want to calculate the mean free path ℓ of an atom in a classical gas, the total cross section σ appearing in the famous formula $\ell \simeq 1/(n\sigma)$ (where n is the density), must be obtained in quantum mechanics (indeed, the attractive van der Waals forces between atoms decay typically as $1/r^7$ at large distances; the potential, in $1/r^6$, thus extends to infinity, and the corresponding *classical* total cross section is thus infinite).

(iii) The concept of cross section is absolutely crucial in modern physics. For example, all collision experiments in nuclear and high-energy physics consist *in fine* in measuring cross sections that can be compared to theoretical models. In the same way, studying the scattering of various particles (photons, either in the visible range or in the X rays range, neutrons, muons ...) allows for the study of the structure of solids and liquids; there also, the concept of cross section is crucial.

(iv) Two different potentials can have the same scattering cross-section; therefore, *inverse scattering problems*, which consist in extracting as much information as possible from the result of scattering experiments, are notoriously difficult and are still the subject of very active research.

Scattering by a hard sphere

The particle being elastically reflected by the sphere, Snell's law implies that the reflection angle is equal to the incidence angle i, which fulfills $\cos i = b/R$ (see Fig. 7.7).

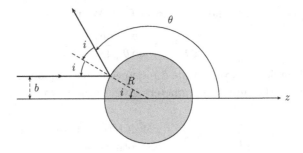

Fig. 7.7 *Scattering of a particle by a hard sphere.*

We thus deduce that the scattering angle $\theta = \pi - 2i$ reads:

$$\theta = \begin{cases} 0 & \text{if } b > R, \\ \pi - 2\arccos(b/R) & \text{otherwise,} \end{cases}$$

or, equivalently, that $b = R\sin(\theta/2)$. Reporting into Eq. (7.2) we get

$$\frac{d\sigma}{d\Omega}(\theta) = \frac{R\sin(\theta/2)}{\sin\theta}\frac{R}{2}\cos(\theta/2) = \frac{R^2}{4}.$$

We observe that the differential cross section does not depend on θ, it is *isotropic*. The calculation of the total cross section is immediate (we just need to multiply by the solid angle corresponding to the whole space, i.e. 4π steradians) and yields

$$\sigma = \pi R^2,$$

which is simply the area of a meridional section of the sphere, as expected. Let us stress that in this specific case of scattering by a hard sphere, the differential cross section does not depend on the collision energy (or, equivalently, on v_∞).

The case of a power-law central force

(1) We now consider a central potential of center O. We then know that the angular momentum with respect to O is a constant, both in direction (implying that the motion is constrained to a plane), and in magnitude; the latter equals, for $t = -\infty$, $L = mv_\infty b$. The total energy E of the particle is also a constant, equal to its value for $t = -\infty$: $E = mv_\infty^2/2$. We use polar coordinates (r, γ) in the plane of the motion. Conservation of angular momentum implies $mr^2\dot{\gamma} = L$, and allows us to eliminate

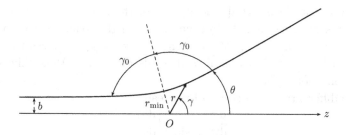

Fig. 7.8   *Scattering of a particle by a repulsive central potential varying as $1/r^n$.*

the angle γ in the expression of the energy E:

$$E = \frac{1}{2}m(\dot{r}^2 + r^2\dot{\gamma}^2) + U(r)$$

$$= \frac{1}{2}m\dot{r}^2 + \frac{L^2}{2mr^2} + U(r). \qquad (7.3)$$

In order to find the scattering angle θ, we eliminate time between $\dot{\gamma} = L/(mr^2)$ and the value of $\dot{r}$ obtained by Eq. (7.3):

$$\frac{d\gamma}{dr} = \frac{\dot{\gamma}}{\dot{r}} = \sqrt{\frac{2}{m}} \frac{L}{mr^2\sqrt{E - L^2/(2mr^2) - \alpha/r^n}}.$$

From Fig. 7.8, we can see that $\theta = \pi - 2\gamma_0$, where γ_0 corresponds to the variation of the polar angle γ when the radius r increases from its minimal value r_{min}, to infinity. We thus conclude that

$$\theta = \pi - 2\int_{r_{min}}^{\infty} \frac{L}{\sqrt{2m}\, r^2\sqrt{E - L^2/(2mr^2) - \alpha/r^n}}\, dr$$

$$= \pi - 2\int_0^{x_{max}} \frac{dx}{\sqrt{1 - x^2 - \beta x^n}}, \qquad (7.4)$$

where the last equality is obtained by the change of variable $x = b/r$, and by using $E = mv_\infty^2/2$ and $L = mv_\infty b$. In Eq. (7.4), we have noted

$$\beta \equiv \frac{2\alpha}{mv_\infty^2 b^n} \qquad (7.5)$$

and $x_{max} = b/r_{min}$ (it is the only non-negative solution of the equation $\beta x^n + x^2 = 1$).

♦ **Remark.**   One can easily check that for $\beta = 0$ (no scattering potential), the scattering angle θ vanishes: we then have $x_{max} = 1$, whence:

$$\theta = \pi - 2\int_0^1 \frac{dx}{\sqrt{1 - x^2}} = \pi - 2\arcsin(1) = 0.$$

(2) Consider a given initial velocity v_∞ of the scattered particle, and a given scattering angle θ. In general, this angle would depend on b and on v_∞; but Eq. (7.4), of the form $\theta = \pi - 2f(\beta)$, shows that for a power-law central potential, θ depends only on β, thus only on the combination $v_\infty^{-2}b^{-n}$.

The differential cross section then reads:

$$\frac{\mathrm{d}\sigma}{\mathrm{d}\Omega} = \frac{b}{\sin\theta}\left|\frac{\mathrm{d}b}{\mathrm{d}\theta}\right|$$

$$= \frac{b}{\sin\theta}\left|\frac{\mathrm{d}b}{\mathrm{d}\beta}\frac{\mathrm{d}\beta}{\mathrm{d}\theta}\right|$$

$$= \frac{b}{\sin\theta}\frac{b}{2n\beta|f'(\beta)|}$$

$$\propto b^2 \quad \text{at fixed } \beta.$$

We have used the fact that Eq. (7.5) implies

$$\frac{\mathrm{d}\beta}{\mathrm{d}b} = -n\frac{\beta}{b}.$$

Consequently, for fixed θ (and thus also fixed β), the differential cross section varies, as a function of v_∞, as:

$$\frac{\mathrm{d}\sigma}{\mathrm{d}\Omega} \propto v_\infty^{-4/n}.$$

It is remarkable that, by simply determining on which (dimensionless) parameter the scattering angle depends, we have been able to find the dependance on v_∞ of the differential cross section, without explicitly calculating the integral appearing in Eq. (7.4). Notice that in the limit $n \to \infty$ (which corresponds to the scattering by a hard sphere[7]) the cross section does not depend on the energy anymore, as was observed in the previous section. Another theoretically interesting case is $n = 4$ (this is called a Maxwellian interaction). In this case, the cross-section varies as $1/v_\infty$, which implies that the rate at which collisions occur in a gas is velocity-independent. This considerably eases the resolution of the Boltzmann kinetic equation, which describes the irreversible dynamics of dilute gases.

(3) For the Coulomb potential $n = 1$, the scattering angle reads

$$\theta = \pi - 2\int_0^{x_{max}}\frac{\mathrm{d}x}{\sqrt{(x_{max} - x)(x + 1/x_{max})}}$$

$$= \pi - 4\arctan x_{max}$$

---

[7]Indeed, if we rewrite the potential as $U(r) = U_0(R/r)^n$, we immediately see that for $n \to \infty$, U approaches 0 (if $r > R$) or ∞ (if $r < R$).

by noticing that $-x^2 - \beta x + 1$ is factored as $(x_{max} - x)(x - 1/x_{max})$ and by using the value of the integral given in the text. We thus deduce

$$x_{max} = \tan\left(\frac{\pi - \theta}{4}\right).$$

Using the well-known formula $\tan x = 2t/(1 - t^2)$, where $t = \tan(x/2)$, we get:

$$\tan\left(\frac{\pi}{2} - \frac{\theta}{2}\right) = \frac{2x_{max}}{1 - x_{max}^2} = \frac{2}{\beta},$$

since, x_{max} and $-1/x_{max}$ being the solutions of $x^2 + \beta x - 1 = 0$, we have $1/x_{max} - x_{max} = \beta$. Finally, since $\tan(\pi/2 - x) = 1/\tan x$, and replacing β by its value, we have:

$$\tan\frac{\theta}{2} = \frac{\alpha}{mv_\infty^2 b}. \tag{7.6}$$

This is the famous Rutherford formula. We now calculate the differential cross section:

$$\frac{d\sigma}{d\Omega} = \frac{b}{\sin\theta}\left|\frac{db}{d\theta}\right|$$

$$= \left(\frac{\alpha}{mv_\infty^2}\right)^2 \frac{1}{\sin\theta\tan(\theta/2)}\left|\frac{d}{d\theta}\frac{\cos(\theta/2)}{\sin(\theta/2)}\right|$$

$$= \left(\frac{\alpha}{2mv_\infty^2}\right)^2 \frac{1}{\sin^4(\theta/2)}. \tag{7.7}$$

We can check that, in agreement with the results of the previous question, $d\sigma \propto v_\infty^{-4}$. The total cross section is infinite, as expected for this long-range potential, since the integral of Eq. (7.7) over $d\Omega = 2\pi\sin\theta d\theta$ diverges at small angles (corresponding, as they should, to large impact parameters). Finally, let us stress that the differential cross section is independent of the sign of α (we have never used here the fact that $\alpha > 0$): thus, the scattering of a negatively charged particle by an atomic nucleus would have the same cross section.

Scattering by a "soft" sphere

(1) First of all, it is obvious that $\theta = 0$ if $b \geqslant R$. We thus consider the case $b < R$, and we choose the origin of time at the moment when the particle reaches the sphere. We will note (z, x) the coordinates of the

particle in the plane defined by O and the initial velocity $\boldsymbol{v}_\infty = v_\infty \boldsymbol{u}_z$. For $t > 0$, and as long as $x^2 + z^2 \leqslant R^2$, the equation of motion reads

$$m\ddot{\boldsymbol{r}} = m\omega^2 \boldsymbol{r},$$

or, upon integration,

$$z(t) = z_0 \cosh(\omega t) + \frac{v_\infty}{\omega} \sinh(\omega t)$$

$$x(t) = b \cosh(\omega t),$$

where we have introduced the notation $z_0 = -\sqrt{R^2 - b^2}$, and taken into account the initial conditions. Defining $t^\star$ as the time when the particle leaves the sphere, the scattering angle fulfills

$$\tan \theta = \frac{v_x(t^\star)}{v_z(t^\star)} = \frac{b\omega}{z_0 \omega + v_\infty \coth(\omega t^\star)}.$$

But $t^\star$ is simply the nonzero solution of the equation $x^2(t^\star) + y^2(t^\star) = R^2$, or, explicitly:

$$\left(R^2 + \frac{v_\infty^2}{\omega^2} \right) \sinh^2(\omega t^\star) = \frac{2v_\infty}{\omega} \sqrt{R^2 - b^2} \sinh(\omega t^\star) \cosh(\omega t^\star).$$

We have used here the fundamental property of hyperbolic trigonometry: $\cosh^2 - \sinh^2 = 1$. Discarding the trivial solution $t^\star = 0$, we find

$$\coth(\omega t^\star) = \frac{v_\infty^2 + R^2 \omega^2}{2v_\infty \omega \sqrt{R^2 - b^2}},$$

which finally allows us to obtain the scattering angle:

$$\tan \theta = \frac{2b\omega^2 \sqrt{R^2 - b^2}}{v_\infty^2 - \omega^2 (R^2 - 2b^2)}. \tag{7.8}$$

(2) In the limit of low-energy scattering $v_\infty \ll \omega R$, the particle cannot penetrate deeply into the sphere, we thus expect to recover the results obtained for the hard sphere. In the limit $v_\infty \ll \omega R$, Eq. (7.8) yields:

$$\tan \theta = \frac{-2b\sqrt{R^2 - b^2}}{R^2 - 2b^2}. \tag{7.9}$$

Let us try to recover this relationship for the scattering by a hard sphere. Starting from the incidence angle i, we have (see Fig. 7.7):

$$\tan \theta = \tan(\pi - 2i)$$

$$= -\tan(2i)$$

$$= -\frac{2\tan i}{1 - \tan^2 i}$$

$$= \frac{-2b/\sqrt{R^2 - b^2}}{1 - b^2/(R^2 - b^2)} \quad \text{since} \quad \tan i = \frac{b}{\sqrt{R^2 - b^2}}$$

$$= \frac{-2b\sqrt{R^2 - b^2}}{R^2 - 2b^2},$$

which indeed corresponds to Eq. (7.9).

(3) In the case of high-energy scattering $v_\infty \gg \omega R$, the particle is only weakly deflected from its initial trajectory. In order to calculate the scattering angle θ, we can therefore assume that the particle has an almost straight trajectory during the interaction with the potential. The transverse velocity imparted to the particle is then approximately

$$v_\perp \simeq \frac{1}{m} \int_{-\Delta t}^{\Delta t} F_x \, dt$$

where the integral is taken over the duration $2\Delta t$ of the interaction with the sphere. The velocity of the particle is almost constant and equal to v_∞, the trajectory is very close to the line of equation $x = b$, and the interaction length is $2\sqrt{R^2 - b^2}$. We conclude that $\Delta t \simeq \sqrt{R^2 - b^2}/v_\infty$. Moreover, the force F_x is almost constant and equal to $m\omega^2 b$. We thus have

$$\theta \simeq \frac{v_\perp}{v_\infty} \simeq \frac{2\Delta t}{m v_\infty} m\omega^2 b,$$

or

$$\theta \simeq \frac{2\omega^2 b \sqrt{R^2 - b^2}}{v_\infty^2}. \tag{7.10}$$

This is indeed the limiting form of Eq. (7.8) for $v_\infty \gg \omega R$.

(4) The value of $x = b/R$ maximizing θ in (7.10) is the one maximizing $x\sqrt{1 - x^2}$; we easily find that the maximum is reached for $x = 1/\sqrt{2}$, and has the value:

$$\theta_{\max} = \frac{\omega^2 R^2}{v_\infty^2}. \tag{7.11}$$

As intuitively expected, the higher the particle velocity, the smaller the deflection.

Application to the atomic structure: the Thomson model and Rutherford's experiment

(1) Let us estimate in a simple way the *order of magnitude* of the size R of a gold atom: the density ρ of gold (in the crystal phase) is approximately equal to the mass of an atom M/N_A divided by the atomic volume, on the order of R^3. We thus find:

$$R \sim \sqrt[3]{\frac{M}{N_A \rho}} = 2.5 \times 10^{-10} \text{ m}.$$

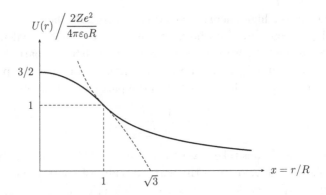

Fig. 7.9 *Interaction energy between the alpha particle and the positive charge distribution of the atom in the Thomson model.*

This is only a very rough number, since we assumed that the crystal is a *compact* assembly of *cubic* (!) atoms of side R. A more accurate calculation, taking into account the fact that a gold crystal is a face-centered cubic (fcc) lattice, gives an atomic radius of about 1.4×10^{-10} m, slightly smaller than the value found above[8].

(2) The electric field created by a uniform, spherical charge distribution of total charge Ze and radius R, is radial: $\boldsymbol{E}(\boldsymbol{r}) = E(r)\boldsymbol{e}_r$. Applying Gauss' theorem on a sphere of radius r, we get:

$$4\pi r^2 E(r) = \frac{Ze}{\varepsilon_0} \times \begin{cases} 1 & \text{if } r \geqslant R; \\ (r/R)^3 & \text{if } r \leqslant R. \end{cases}$$

The electrostatic potential $\mathcal{V}(\boldsymbol{r})$, defined by $\boldsymbol{E} = -\boldsymbol{\nabla}\mathcal{V}$, is obtained by integrating the field E. We chose $\mathcal{V}(\infty) = 0$; the continuity of the potential in $r = R$ (for a volumic charge distribution, there is no discontinuity in the electrostatic potential) allows us to fix the integration constant for $r \leqslant R$. We finally have:

$$\mathcal{V}(r) = \frac{Ze}{4\pi\varepsilon_0} \times \begin{cases} 1/r & \text{if } r \geqslant R; \\ (3 - r^2/R^2)/(2R) & \text{if } r \leqslant R. \end{cases}$$

The interaction potential energy between the alpha particle and the positive charge distribution simply reads $U(r) = 2e\mathcal{V}(r)$ and is shown in Fig. 7.9.

Numerically, we find, for $R = 10^{-10}$ m, that the maximum of potential energy is $U(0) \simeq 3.4$ keV.

---

[8]To get this numerical value, one needs to know — or calculate — the *compacity* of a fcc crystal, which is equal to $\pi/\sqrt{18} \simeq 0.74$.

(3) In order to study the scattering of alpha particles, we can neglect the recoil of the gold nucleus because its mass $M \simeq 192m_{\mathrm{p}}$, where m_{p} is the proton mass, is much higher than that of the alpha particle ($\simeq 4m_{\mathrm{p}}$): the reduced mass in the two-body problem is almost equal to the one of the alpha particle, which almost coincides with the "fictitious" particle. We can also in a first approximation neglect the effect of the electrons, because they have a very small mass, and thus they almost do not perturb the motion of the alpha particle. However, the electrons screen the potential created by the positive charges, which leads to a decrease of the deflection; but, as we shall see, even by considering only positive charges, the calculated deflections are very weak for an extended charge distribution.

(4) We assume that the Thomson model is valid. If the alpha particle does not enter into the charge distribution ($r(t) > R$ for all times), we are in the case of the scattering by the Coulomb potential, as in this case $U(r) \propto 1/r$. The deflection is maximal for the minimal impact parameter b such that at all times $r(t) > R$, i.e. such as $r_{\min} > R$. We obtain $r_{\min}$ by equating to zero the radial velocity $\dot{r}$ in

$$E = \frac{mv_\infty^2}{2} = \frac{m\dot{r}^2}{2} + \frac{L^2}{2mr^2} + \frac{\alpha}{r},$$

with $L = mv_\infty b$, and we get:

$$r_{\min} = \frac{\alpha}{mv_\infty^2} + \sqrt{\left(\frac{\alpha}{mv_\infty^2}\right)^2 + b^2}. \tag{7.12}$$

We conclude that the impact parameter $b_{\min}$ corresponding to a particle which barely touches the charge distribution is

$$b_{\min} = \sqrt{R^2 - \frac{2\alpha R}{mv_\infty^2}},$$

and, by substituting into Eq. (7.6), we have:

$$\tan\left(\frac{\theta_{\max}}{2}\right) = \frac{\alpha}{mv_\infty^2 b_{\min}} \simeq 2 \times 10^{-4}.$$

For the numerical values, we use $\alpha = 2Ze^2/(4\pi\varepsilon_0)$, and the kinetic energy of the alpha particles $E = mv_\infty^2/2 = 5.6$ MeV, which allows us to calculate $v_\infty \simeq 1.6 \times 10^7$ m $\cdot$ s^{-1}, showing incidently that the non-relativistic approximation we used is valid since $v \ll c$. The maximum deflection is thus smaller than 10^{-3}, and collisions with such a given impact parameter cannot explain the observation of highly deflected ($\theta \sim \pi/2$) particles.

(5) In order to try to explain the observed large deflections, we are thus led to consider that the strongly deflected alpha particles enter into the extended charge distribution. Since $E = 5.6$ MeV $\gg U(0) = 3.4$ keV, we are in the case of high-energy scattering, and we can thus use the result (7.11) to find the maximal deflection. By writing that $3m\omega^2 R^2/2 = U(0)$ (the factor of 3 comes from the fact that the soft sphere giving a potential identical, for $r < R$, to the one of the actual distribution, has a radius $\sqrt{3}R$, see Fig. 7.9), we obtain:

$$\theta_{max} = \frac{U(0)}{3E} \simeq 2 \times 10^{-4}.$$

Thus, if the positive charge was extended over the whole volume of an atom, the deflection of alpha particles with such a high energy *could not be* as large as what is observed experimentally.

(6) In the framework of Rutherford's model, the minimal distance between the alpha particle and the positive charges is larger than the size of this distribution; we thus deal with the case of Coulomb scattering. We therefore look for the minimum distance of approach for the scattering under an angle $\theta = \pi/2$, which means, using Eq. (7.6), that the impact parameter reads:

$$b = \frac{\alpha}{mv_\infty^2}.$$

By substitution into Eq. (7.12), we obtain as the minimal distance:

$$r_{min} = \frac{\alpha}{mv_\infty^2}(1 + \sqrt{2}).$$

Numerically we find $r_{min} \simeq 5 \times 10^{-14}$ m; this is an upper bound on the size of the atomic nucleus. It is now known that the typical size of atomic nuclei is 10^{-15} m. Note that by using more energetic particles, we can "push down" this upper bound, this is, roughly speaking, the reason why particle accelerators have higher and higher energies (see Problem 7.2).

♦ **Remark.** One might think that in order to study Rutherford scattering, i.e. a microscopic phenomenon, quantum mechanics is necessary. In fact, one can show that the classical result (7.7) for the scattering cross section is still valid quantum mechanically (for purely fortuitous reasons).

▶ **Further reading.** The original articles by Geiger and Marsden on the one hand, and by Rutherford on the other hand, are freely available on the internet:

- H. Geiger and E. Marsden, *On a diffuse reflection of the α-particles*, Proc. Roy. Soc. **82A**, 495 (1909), available at http://rspa.royalsocietypublishing.org/content/82/557/495;
- E. Rutherford, *The scattering of α and β particles by matter and the structure of the atom*, Philosophical Magazine, **6(21)**, 669 (1911), available at http://www.math.ubc.ca/~cass/rutherford/rutherford.html.

7.5 Slowing atoms with a moving wall **

We describe in this exercise the one-dimensional propagation of a bunch of N atoms without interactions. All the atoms are initially located at the same point $z = 0$. Their velocity is distributed according to the Gaussian probability distribution

$$P(v_z) = \frac{1}{\sqrt{2\pi}\Delta}e^{-(v_z-\bar{v})^2/(2\Delta^2)}.$$

Free propagation

(1) Comment on the expression for the probability distribution $P(v_z)$.
(2) Calculate the linear density $n(z,t)$ by evaluating the number of particles lying between z and $z + \mathrm{d}z$ at a given time t.

Collision with a potential wall moving at a constant velocity

In this section, we consider (see Fig. 7.10) that the atom bunch is going towards an infinite potential wall moving at a velocity V, and that was initially located at $z = d$. Each atom of the bunch undergoes an elastic collision on this "moving mirror".

(1) Determine, for a particle with an initial velocity v_z, the time t^* and position z^* of its collision with the moving potential wall. What is the velocity of the particle after its collision with the wall?
(2) Draw in the phase space plane (z, v_z) the curve $v_z(z^*)$. Show graphically the velocity of the particle as a function of the position after its collision with the wall.
(3) Determine the expression of the linear atomic density $n(z,t)$ by writing down the local balance of the number of particles lying in between z and $z + \mathrm{d}z$ at time t when the bunch is undergoing a collision with the wall.

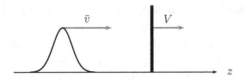

Fig. 7.10   *A bunch of atoms is heading towards the moving potential wall.*

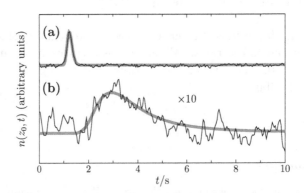

Fig. 7.11   *Linear atomic density measured at $z_0 = 1.75$ m for (a) a freely propagating bunch, and for (b) a bunch after its interaction with the moving potential wall (d = 0.5 m). The noisy, thin lines correspond to the experimental data, the gray lines are the results of a fit according to the theoretical prediction.*

(4) Figure 7.11 shows experimental measurements[9] of the linear density $n(z_0, t)$ of bunch with typically $N \simeq 10^9$ "cold" atoms, measured as a function of time at $z_0 = 1.75$ m: in (a) for a free propagation (without the moving mirror), in (b) after interaction with the moving mirror initially at $d = 0.5$ m. Estimate the mean velocity $\bar{v}$ of the bunch, its velocity dispersion Δ, the mirror velocity V and the mean velocity $v_{\rm f}$ of the bunch after its interaction with the moving potential wall.

(5) What is the variation of kinetic energy of the atom bunch which results from the collision, using the data of Fig. 7.11?

---

[9]G. Reinaudi, Z. Wang, A. Couvert, T. Lahaye and D. Guéry-Odelin, *A moving magnetic mirror to slow down a bunch of atoms*, Eur. Phys. J. D **40**, 405 (2006) (also freely available on the internet at http://arxiv.org/abs/cond-mat/0608710).

Stopping atoms of arbitrary velocities with an accelerated wall

Let us assume now that a bunch of particles is released at $z = 0$ at time $t = 0$ with an arbitrary velocity distribution (the only restriction being that it is nonzero only for positive velocities). Show that there exists a trajectory $z_\mathrm{M}(t)$ of the mirror, such that all particles starting at the origin and moving with positive velocity will be stopped by their collision with the wall, independently of their initial velocity. We assume that the wall starts moving from $z = 0$ at $t = 0$.

Solution

Free propagation

(1) One has a Gaussian distribution centered on the mean velocity $\bar{v}$, and characterized by the velocity dispersion Δ. Note that a bunch at thermal equilibrium at temperature T and with a mean velocity $\bar{v}$ would have exactly the same velocity distribution along the z axis, with a velocity dispersion $\Delta = \sqrt{k_\mathrm{B}T/m}$, where m denotes the mass of an atom.

(2) At a given time t, the number of particles $\mathrm{d}N(z,t)$ in between z and $z + \mathrm{d}z$ is given by the initial population of atoms with a velocity v_z and $v_z + \mathrm{d}v_z$, where $z = v_z t$ since the propagation is free:

$$\mathrm{d}N(z) = NP(v_z)\mathrm{d}v_z = \frac{N}{t}P\left(\frac{z}{t}\right)\mathrm{d}z.$$

We infer the expression for the linear atomic density

$$n(z,t) = \frac{\mathrm{d}N(z,t)}{\mathrm{d}z} = \frac{N}{t}P\left(\frac{z}{t}\right).$$

At a given time t, one can easily check the normalization of the linear atomic density by integrating it over the z axis:

$$\int_{-\infty}^{\infty} n(z,t)\mathrm{d}z = \int_{-\infty}^{\infty} \frac{N}{t}P\left(\frac{z}{t}\right)\mathrm{d}z = N,$$

where the last equality is obtained by using the variable $v_z = z/t$.

Collision with a moving potential wall

(1) At time t, the particle, with velocity v_z, is located at $v_z t$, and the wall is at $d + Vt$. The collision thus occurs at $z^\star$ after a time $t^\star$ such that

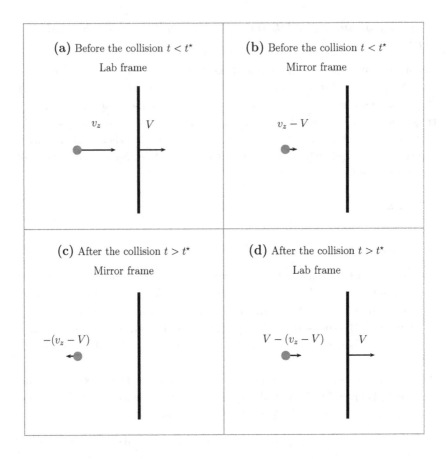

Fig. 7.12   *Representation of the velocities of the particle and the mirror before the collision ($t < t^*$) in the laboratory frame (a), in the mirror frame (b), after the collision ($t > t^*$) in the mirror frame (c), and in the laboratory frame (d).*

$z^* = v_z t^* = d + V t^*$, where V is the velocity of the moving wall. We thus obtain

$$t^* = \frac{d}{v_z - V}, \quad z^* = \frac{v_z}{v_z - V} d, \quad \text{and} \quad v_z(z^*) = \frac{V z^*}{z^* - d}.$$

In the inertial frame attached to the potential wall, the velocity of the particle changes sign after the elastic collision with the mirror. The velocity changes from $(v_z - V)$ to $-(v_z - V)$ in this frame, so that v_z is changed to $V - (v_z - V) = 2V - v_z$ in the laboratory frame (see Fig. 7.12).

(2) The curve $v_z(z^*)$ is a hyperbola (depicted in Fig. 7.13 as a thick solid

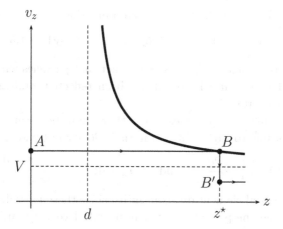

Fig. 7.13 *Representation of the "elastic collision" of a particle with an infinite potential wall moving at velocity V.*

line), with the lines $z = d$ and $v_z = V$ as asymptotes. A particle of velocity v_z (point A) is moving along the line segment AB during its free propagation. The collision with the moving mirror occurs at z^* at B. Just after the collision, the point representing the trajectory is at B', symmetric of point B with respect to the horizontal line of fixed velocity V. After the collision, the atom evolves on a straight line of ordinate $(2V - v_z)$, i.e. the one starting at B'. The position, after the collision, is therefore given by:

$$z = z^* + (2V - v_z)(t - t^*) \quad \text{for} \quad t > t^*,$$

or, using the expression for t^* and z^*, by

$$z = 2d + (2V - v_z)t. \tag{7.13}$$

(3) The linear density contains the contribution of two types of particles, those that have not yet been reflected (nr) and those that have already undergone the collision with the moving potential wall (r):

$$n(z,t) = \left.\frac{dN(z,t)}{dz}\right|_{\text{nr}} + \left.\frac{dN(z,t)}{dz}\right|_{\text{r}}.$$

At a given time t, only the particles with an initial velocity larger than $V + d/t$ have been reflected. The distribution for the non-reflected particles is thus given by:

$$\left.\frac{dN(z,t)}{dz}\right|_{\text{nr}} = \frac{N}{t} P\left(\frac{z}{t}\right) \Theta(Vt + d - z),$$

where $\Theta(x)$ denotes the Heaviside step function

$$\Theta(x) = 1 \quad \text{if} \quad x > 0, \quad \Theta(x) = 0 \quad \text{otherwise.}$$

The Heaviside function expresses the fact that particles with a velocity smaller than $V + d/t$ have not yet been reflected, and are therefore before the mirror.

For the reflected particles, those that are in between z and $z + dz$ had, according to Eq. (7.13), an initial velocity in between

$$v_z = 2V - \frac{z}{t} + \frac{2d}{t} \quad \text{and} \quad v_z + dv_z = 2V - \frac{z}{t} - \frac{dz}{t} + \frac{2d}{t}.$$

The differential element dv_z is negative since those particles have been reflected and the particles at z had an initial velocity larger than the particles at $z + dz$ if $dz > 0$. We deduce the number of reflected atoms in the "slice" dz:

$$dN_r(z) = NP(v_z)dv_z = \frac{N}{t} P\left(2V + \frac{2d}{t} - \frac{z}{t}\right) \Theta(Vt + d - z)dz.$$

The factor $\Theta(Vt + d - z)$ simply comes from the fact that all particles that have been reflected are located before the mirror. We finally obtain the expression for the linear atomic density by adding the two contributions:

$$n(z,t) = N \frac{\Theta(Vt + d - z)}{t} \left[P\left(\frac{z}{t}\right) + P\left(2V + \frac{2d}{t} - \frac{z}{t}\right)\right]. \quad (7.14)$$

At a given time t, one can, as before, check the normalization of the linear atomic density by integrating over the z axis:

$$\int_{-\infty}^{\infty} N \frac{\Theta(Vt + d - z)}{t} \left[P\left(\frac{z}{t}\right) + P\left(2V + \frac{2d}{t} - \frac{z}{t}\right)\right] dz = N.$$

(4) From Fig. 7.11(a), one can deduce approximately the initial velocity of the bunch and its velocity dispersion. The maximum of the bunch is located at time $\tau \simeq 1.2$ s, and it has propagated over a distance $z_0 = 1.75$ m during this time; its mean initial velocity is therefore $\bar{v} = z_0/\tau \simeq 1.75/1.2 = 1.45 \, \text{m} \cdot \text{s}^{-1}$. Its full width is on the order of 0.3 s for a reduction of the peak size by a factor on the order of ~ 7 (i.e. at $1/e^2$). To deduce the velocity dispersion, we use the following rough estimate[10]. The particles with a velocity $\bar{v} \pm \Delta$ need a time $z_0/(\bar{v} \pm \Delta)$

---

[10] An accurate determination is made by fitting the experimental data with the expected theoretical curve (7.14): see the gray solid lines in Fig. 7.11.

to reach the detection zone. The full width Δt of the experimental signal is therefore on the order of

$$\Delta t \simeq \frac{z_0}{\bar{v} - \Delta} - \frac{z_0}{\bar{v} + \Delta} = \frac{2z_0\Delta}{\bar{v}^2 - \Delta^2} \simeq \frac{2z_0\Delta}{\bar{v}^2}.$$

We deduce that $\Delta \simeq \bar{v}\Delta t/(2\tau) = 0.18\,\text{m} \cdot \text{s}^{-1}$. Note that this corresponds to a temperature $T = m\Delta^2/k_B \simeq 340\,\mu\text{K}$ (assuming that we deal with rubidium atoms, with mass $m = 1.45 \times 10^{-25}$ kg). Such a low temperature can be obtained by laser cooling techniques (see the problems of Chapter 9).

The signal obtained for the reflected bunch (i.e. when all particles of the initial bunch have been reflected, even the slower ones, see Fig. 7.11(b)) has its maximum at $t_0 \simeq 2.8$ s at $z_0 = 1.75$ m when the argument of the corresponding distribution P vanishes:

$$2V + \frac{2d}{t_0} - \frac{z_P}{t_0} = \bar{v} \quad \Longrightarrow \quad V \simeq 0.85\,\text{m} \cdot \text{s}^{-1}.$$

We infer the mean velocity of the bunch $v_f = 2V - \bar{v} \simeq 0.25\,\text{m} \cdot \text{s}^{-1}$.

(5) Using König's theorem (see the reminder at the end of the solution), the kinetic energy of an ensemble of atoms can be rewritten as the contribution of the center of mass of the bunch and of its velocity dispersion:

$$E_k = \sum_{i=1}^{N} \frac{1}{2}mv_i^2 = \sum_{i=1}^{N} \frac{1}{2}m(v_i^2 - \bar{v}^2) + \sum_{i=1}^{N} \frac{1}{2}m\bar{v}^2$$

from which one finds:

$$E_k = \frac{Nm}{2}\left(\Delta^2 + v^2\right). \tag{7.15}$$

Neglecting the velocity dispersion, the fraction η_1 of kinetic energy (corresponding to the mean velocities) that has been absorbed by the potential wall is given by:

$$\eta_1 = 1 - \frac{v_f^2}{\bar{v}^2} \simeq 0.97.$$

The collision with the moving potential wall yields a reduction by 97% of the initial kinetic energy of the bunch! One can show that the velocity dispersion is essentially unaffected in the collision process, the actual fraction η of absorbed kinetic energy is therefore given by:

$$\eta = 1 - \frac{v_f^2 + \Delta^2}{\bar{v}^2 + \Delta^2} \simeq 0.96.$$

In this specific example, the velocity dispersion plays a negligible role. One may wonder what happens with the momentum of the particles. The answer is that the potential wall absorbs the exchanged momentum. As the mass of the wall is assumed to be infinite, its recoil is negligible. This assumption is reasonable: indeed, the mass of the reflected atoms, the number of which is on the order of one billion, is extremely small in comparison to the one of the macroscopic wall. We emphasize that a similar technique can be used to slow down neutrons beams, and was recently applied to the slowing of a supersonic beam of helium atoms[11].

Stopping atoms of arbitrary velocities with an accelerated wall

We consider a particle emitted at $z = 0$ and $t = 0$ with an unknown velocity $v_z \geq 0$. According to the first part of the exercise, the particle is at rest in the laboratory frame after the collision with the wall if the mirror velocity, $V(t^*)$, has a velocity equal to half the velocity of the particle at the time t^* of the collision. This criterium reads

$$\frac{dz_{\mathrm{M}}}{dt}\bigg|_{t^*} = \frac{v_z}{2}.$$

Since the particle evolves freely, the collision occurs at $z_{\mathrm{M}}(t^*) = v_z t^*$. We therefore obtain the equation

$$\frac{dz_{\mathrm{M}}}{dt}\bigg|_{t^*} = \frac{z_{\mathrm{M}}(t^*)}{2t^*}. \tag{7.16}$$

This equation has to be valid for any positive value of v_z and therefore for any time t^*. The trajectory of the mirror is thus the solution of the ordinary differential equation (7.16) for the variable t^* and with the initial condition $z_{\mathrm{M}}(0) = 0$. We thus find:

$$z_{\mathrm{M}}(t) = \alpha\sqrt{t},$$

where the constant $\alpha > 0$ is, in principle, arbitrary and determines the position and time at which the collision occurs for a given particle of initial velocity $v_z > 0$. We emphasize that this is a remarkable and nonintuitive result.

---

[11]E. Narevicius, A. Libson, M.F. Riedel, C.G. Parthey, I. Chavez, U. Even, and M.G. Raizen, *Coherent Slowing of a Supersonic Beam with an Atomic Paddle*, Phys. Rev. Lett. **98**, 103201 (2007), also available on the web at http://arxiv.org/abs/physics/0701003.

► **Further reading.** The reader is invited to read the original article: S. Schmidt, J.G. Muga, and A. Ruschhaupt, *Stopping particles of arbitrary velocities with an accelerated wall*, Phys. Rev. A **80**, 023406 (2009), also freely available on the internet at the address http://arxiv.org/abs/0903.0071.

Reminder: König's theorem

König's theorem states that the kinetic energy E_k of a system of particles of masses m_i and velocities v_i can be calculated as the sum of:

- the kinetic energy

$$E_{com} = \frac{1}{2}MV^2$$

  associated with the motion of the center of mass, i.e. the kinetic energy of a particle of mass $M = \sum_i m_i$ equal to the total mass of the system and having the velocity $V = \sum m_i v_i / M$ of the center of mass;

- the kinetic energy

$$E_k^\star = \sum_i \frac{1}{2}m_i(v_i - V)^2$$

  of the particles in the center of mass frame.

The proof is as follows. By writing the trivial identity $v_i^2 = (v_i - V + V)^2$ and expanding the square, one has

$$E_k = \sum_i \frac{1}{2}m_i(v_i - V)^2 + \sum_i m_i V \cdot (v_i - V) + \frac{1}{2}\left(\sum_i m_i\right)V^2$$

$$= E_k^\star + V \cdot \left(\sum_i m_i v_i - MV\right) + E_{com}.$$

Since the terms between brackets vanish by definition of V, one directly obtains König's theorem.

Chapter 8

Manipulation of Charged Particles

Since the seventies, physicists have known how to trap charged particles (electrons and ions) using electromagnetic fields. For instance, it is now possible to trap a single electron in a Penning trap and to keep it for months! There are many applications, either in metrology (one can perform extremely accurate measurements of mass, mass ratios, gyromagnetic factors ...), or more recently in the field of quantum information processing. The Nobel prize in Physics was awarded in 1989 to Wolfgang Paul and Hans Dehmelt[1] for their pioneering work in the trapping of charged particles. In the problems of this chapter, a certain number of techniques allowing for the manipulation of ions and electrons are studied.

8.1 Coulomb crystal *

By using ion traps (for instance Paul traps, see Problem 8.4), it is nowadays possible to confine a large number of ions in harmonic traps. When one cools these ions (using laser cooling techniques, see Problem 9.3), they organize into an ordered state, called a Coulomb crystal. We will show, describing the ion assembly in the continuum limit, that the ionic density is constant in such a harmonically confined crystal.

(1) We consider ions of charge q and mass m, confined in an isotropic, three-dimensional harmonic trap, with angular frequency ω. For symmetry reasons, the ion density $n(r)$ depends only on the radial coordinate r. Calculate the electric field created at any point in space by the charge

---

[1]The third laureate of 1989, Norman Ramsey, was awarded the prize for the invention of a spectroscopic technique that has allowed for a considerable improvement of the accuracy of atomic clocks.

distribution.

(2) Writing the condition of mechanical equilibrium for the ensemble of ions, show that the density is constant with the value:

$$n = \frac{3m\omega^2\varepsilon_0}{q^2}.$$

(3) The photograph reproduced in Fig. 8.1 displays a cut through such a Coulomb crystal obtained for a trap of frequency $\omega/(2\pi) = 600$ kHz, with $^{40}Ca^+$ ions[2]. Comment.

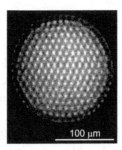

100 μm

Fig. 8.1 *Coulomb crystal containing 2,700 ions* $^{40}Ca^+$. *Figure reproduced (with permission) from the article Phys. Rev. Lett.* **96**, *103001 (2006). Copyright (2006) American Physical Society.*

Solution

(1) For symmetry reasons, the electric field E is directed along the radial direction, and its magnitude $E(r)$ depends only on r. By using the Gauss theorem on a sphere of radius r, we have

$$4\pi r^2 E(r) = \frac{Q_{\text{int}}}{\varepsilon_0} = \frac{1}{\varepsilon_0}\int_0^r 4\pi x^2 qn(x)\,\mathrm{d}x.$$

Upon differentiating with respect to r, we obtain the following relationship between the electric field and the ionic density:

$$\frac{\mathrm{d}}{\mathrm{d}r}\left[r^2 E(r)\right] = \frac{q}{\varepsilon_0}r^2 n(r). \tag{8.1}$$

[2]A. Mortensen, E. Nielsen, T. Matthey, and M. Drewsen, *Observation of three-dimensional long-range order in small ion Coulomb crystals in an rf trap*, Phys. Rev. Lett. **96**, 103001 (2006). This article is also freely available on the internet at http://arxiv.org/abs/physics/0508125.

(2) The charge $dQ = 4q\pi r^2 n(r) \, dr$ (with mass $dM = 4m\pi r^2 n(r) \, dr$) contained inside the spherical shell located between r and $r + dr$ is in equilibrium when the harmonic restoring force $-dM\omega^2 r u_r$ is compensated for by the electrostatic repulsion $dQE(r)u_r$. We thus get, after simplifying by $4\pi r^2 n(r)$:

$$E(r) = \frac{m\omega^2}{q} r.$$

Substituting into Eq. (8.1), this finally yields:

$$n(r) = \frac{3m\omega^2 \varepsilon_0}{q^2}. \tag{8.2}$$

The ionic density is thus constant, up to a radius R such as $4\pi R^3 n/3 = N$, where N is the total number of ions in the crystal. Note that, as intuitively expected, the density increases if one increases the confinement ω, and decreases if q increases, for in this case the repulsion between ions becomes stronger.

(3) The photograph allows us to see that, indeed, the crystal density seems to be a constant, qualitatively at least: the mean distance between two nearest-neighbor ions is the same for ions close to the centre or for ions close to the boundary of the crystal. In order to estimate the density of this crystal, one can use two methods. The first one consists in using the information giving the total ion number $N = 2,700$ in the crystal, and in measuring, using the scale shown on the image, its radius R. We find $R \simeq 90 \ \mu m$. The ion density is then obtained simply by solving $4\pi R^3 n/3 = N$, yielding $n \simeq 10^{15} \ \mathrm{m}^{-3}$. An alternative method consists in measuring the mean distance ℓ between ions (one counts about ten ions on a $100 \ \mu m$ long line, implying $\ell \simeq 10^{-5} \ \mathrm{m}$). We can then estimate easily the density as $n \simeq \ell^{-1/3}$, giving of course the same result as before. It is actually in this way that the total number N of ions in the crystal is measured.

Let us compare this experimental value with our theoretical estimate. For $^{40}\mathrm{Ca}^+$ ions, we have $m \simeq 40 m_\mathrm{p} \simeq 6.7 \times 10^{-26}$ kg (m_p being the proton mass); the charge is equal to the elementary one: $q = 1.6 \times 10^{-19}$ C. Substituting those numbers into Eq. (8.2) we get $n \simeq 10^{15} \ \mathrm{m}^{-3}$, in very good agreement with the measurements.

♦ **Remark.** Another system that exhibits this type of competition between a harmonic confinement and a $1/r^2$ repulsive force between the particles is a trap for neutral atoms, the *magneto-optical trap*[3]. In this type of trap, a combination of laser beams and of inhomogeneous magnetic fields allows for the confinement of atoms (and for their cooling, down to very low temperatures, on the order of 10^{-4} K, by cooling mechanisms similar to Doppler cooling[4]). In spite of the fact that the atoms are neutral, and thus do not experience Coulomb repulsion, these atoms experience repulsive forces because they keep absorbing photons emitted by other atoms contained in the trap (photons that were first absorbed in the trapping beams). But the photon flux emitted by an atom decays as $1/r^2$, where r is the distance from the emitting atom, and the radiation pressure force thus also decreases as $1/r^2$. Within this simple model, the atomic density in a magneto-optical trap is thus expected to be constant[5].

8.2   Normal modes of a one-dimensional ion chain ⋆⋆

For two to three decades now, physicists have known how to realize ion traps (for instance linear Paul traps) that allow for the confinement of particles of mass m and charge q, whose motion is restricted along a direction x. We consider N identical ions, which are all subjected to an external trapping potential $U(x) = m\omega^2 x^2/2$, and which also experience the electrostatic repulsion exerted by all other ions. In the following, we will use the notation:

$$e^2 \equiv \frac{q^2}{4\pi\varepsilon_0}.$$

Case of two ions: N = 2

(1) What are the equilibrium positions $x_{1,e}$ and $x_{2,e}$ of the two ions?
(2) Study the small oscillations around equilibrium: give the eigenfrequencies of the normal modes, and describe what they look like.

Case of three ions: N = 3

(1) What are the equilibrium positions $x_{i,e}$ $(i = 1, 2, 3)$ of the three ions?

---

[3] Whose working principle was proposed by the physicist Jean Dalibard, and which was operated for the first time in Steven Chu's group (Steven Chu obtained the Nobel Prize in Physics in 1997 together with Claude Cohen-Tannoudji and William Phillips).

[4] See Problem 9.3.

[5] For more details, see e.g. T. Walker, D. Sesko, and C.E. Wieman, *Collective behavior of optically trapped neutral atoms*, Phys. Rev. Lett. **64**, 408 (1990).

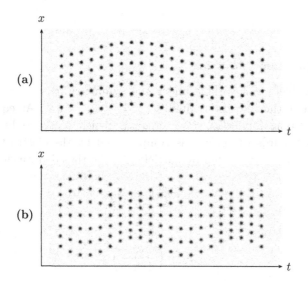

Fig. 8.2 *Normal modes of a chain containing seven ions: in (a) one can see the dipole (sloshing) mode; in (b) the monopole (breathing) mode. The time between successive images is the same for both images. Figure adapted from H.-C. Nagerl's PhD thesis, available on the internet at http://heart-c704.uibk.ac.at/publications/dissertation/hcn_diss.pdf.*

(2) Study the small oscillations around equilibrium: give the eigenfrequencies of the normal modes, and describe what they look like. Comment.

An experimental example

Figure 8.2 shows two series of images, taken at regular time intervals, of a chain of seven calcium ions (the chain is vertical on the figure) irradiated by laser beams, and whose fluorescence is imaged. In (a), the oscillation of the center of mass (dipole mode, or sloshing mode) is excited; in (b), the breathing mode (monopole mode) is excited. Show that the ratio of the frequencies of those two modes (which can be shown to be the lowest frequency modes) that are experimentally observed agrees with the calculations above (one can show that this ratio is independent of N, see at the end of the solution).

Solution

Case of two ions: $N = 2$

(1) The problem is invariant under the symmetry $x \to -x$; therefore the equilibrium positions fulfill $x_{1,e} = -x_{2,e} = x_0$. At equilibrium, the trapping force $F_{trap} = -m\omega^2 x_0$, which tends to bring particle (1) towards the origin, is compensated by the electrostatic force $F_{rep} = e^2/(2x_0)^2$ exerted by particle (2). We thus deduce that

$$m\omega^2 x_0 = \frac{e^2}{4x_0^2},$$

or:

$$x_0 = \left(\frac{e^2}{4m\omega^2}\right)^{1/3}$$

In agreement with intuition, the distance $2x_0$ between the two ions at equilibrium increases with e^2 (i.e. with electrostatic repulsion), as well as when ω decreases (i.e. when the trapping becomes weaker).

(2) Newton's second law, applied to ions (1) and (2), reads:

$$m\ddot{x}_1 = -m\omega^2 x_1 + \frac{e^2}{(x_1 - x_2)^2} \tag{8.3}$$

$$m\ddot{x}_2 = -m\omega^2 x_2 - \frac{e^2}{(x_1 - x_2)^2} \tag{8.4}$$

By adding Eqs (8.3) and (8.4) we obtain, for the variable $X \equiv (x_1 + x_2)/2$, the equation

$$\ddot{X} = -\omega^2 X.$$

The position X of the center of mass thus undergoes harmonic oscillations with angular frequency ω, independent of the Coulomb interactions[6]. We have thus found a first normal mode of the ion chain, the mode of the center of mass[7], which oscillates at the trap frequency ω. The two ions are moving in phase, their mutual distance remaining equal to the equilibrium one.

---

[6]We could have obtained this result immediately by applying Newton's second law directly to the system consisting of both ions, for the Coulomb forces are then *internal* forces.

[7]Also called, depending on the context, the dipole mode, the sloshing mode, or the Kohn mode.

It is clear for symmetry reasons that the other normal mode corresponds to out-of-phase oscillations of the two ions. We thus write $x_1(t) = x_0 + \varepsilon(t)$, $x_2(t) = x_0 - \varepsilon(t)$, where $\varepsilon \ll x_0$, and we substitute into Eq. (8.3) to obtain:

$$m\ddot{\varepsilon} = -m\omega^2 x_0 - m\omega^2 \varepsilon + \frac{e^2}{4(x_0 + \varepsilon)^2}$$

$$\simeq -m\omega^2 x_0 - m\omega^2 \varepsilon + \frac{e^2}{4x_0^2}\left(1 - 2\frac{\varepsilon}{x_0}\right)$$

$$\simeq -m\omega^2 \varepsilon - \frac{e^2}{2x_0^2}\frac{\varepsilon}{x_0}$$

where we kept only the terms of first order in ε/x_0, and used the value of x_0 to cancel out the zeroth-order term. Using again the value of x_0, we finally get:

$$\ddot{\varepsilon} = -3\omega^2 \varepsilon \,.$$

This oscillation mode, where the two ions oscillate out-of-phase, with the center of mass standing still, is called the breathing mode[8], and has a frequency $\sqrt{3}\omega$.

For arbitrary initial conditions, the motion of the ions is a superposition of the monopole and dipole modes.

Case of three ions: $N = 3$

(1) It is obvious, for symmetry reasons, that, at equilibrium, the central ion is in $x_{2,e} = 0$, with the two other ions symmetrically located: $x_{1,e} = -x_{3,e} = \tilde{x}_0$. In order to find $\tilde{x}_0$, we write the equilibrium condition for ion (1):

$$-m\omega^2 \tilde{x}_0 + e^2 \left(\frac{1}{\tilde{x}_0^2} + \frac{1}{4\tilde{x}_0^2}\right) = 0 \,.$$

We thus immediately deduce that:

$$\tilde{x}_0 = 5^{1/3} x_0 \,.$$

(2) Let us write down the equations of motion for the positions $x_i(t)$ ($i = $

---

[8]Or monopole mode.

1, 2, 3) of the ions. Newton's second law gives

$$\ddot{x}_1 = -\omega^2 x_1 + \frac{e^2}{m}\left(\frac{1}{(x_2-x_1)^2} + \frac{1}{(x_3-x_1)^2}\right), \tag{8.5}$$

$$\ddot{x}_2 = -\omega^2 x_2 + \frac{e^2}{m}\left(\frac{1}{(x_3-x_2)^2} - \frac{1}{(x_1-x_2)^2}\right), \tag{8.6}$$

$$\ddot{x}_3 = -\omega^2 x_3 - \frac{e^2}{m}\left(\frac{1}{(x_2-x_3)^2} + \frac{1}{(x_1-x_3)^2}\right). \tag{8.7}$$

We note $x_1 = \tilde{x}_0 + \varepsilon_1$, $x_2 = \varepsilon_2$, and $x_3 = -\tilde{x}_0 + \varepsilon_3$, where ε_i are first order infinitesimals. By performing the expansion

$$\frac{1}{(x_1-x_2)^2} = \frac{1}{(\tilde{x}_0 + \varepsilon_1 - \varepsilon_2)^2} \simeq \frac{1}{\tilde{x}_0^2}\left(1 - 2\frac{\varepsilon_1 - \varepsilon_2}{\tilde{x}_0}\right)$$

(and similarly for $(x_2-x_3)^{-2}$ and $(x_1-x_3)^{-2}$), and then substituting into Eqs (8.5) to (8.7), we get:

$$\ddot{\varepsilon}_1 = -\frac{\omega^2}{5}\left(14\varepsilon_1 - 8\varepsilon_2 - \varepsilon_3\right),$$

$$\ddot{\varepsilon}_2 = -\frac{\omega^2}{5}\left(-8\varepsilon_1 + 21\varepsilon_2 - 8\varepsilon_3\right),$$

$$\ddot{\varepsilon}_3 = -\frac{\omega^2}{5}\left(-\varepsilon_1 - 8\varepsilon_2 + 14\varepsilon_3\right),$$

where we made use of the value of $\tilde{x}_0$ calculated in the previous question. We look for solutions of the form $\varepsilon_{i,0}e^{i\lambda\omega t}$, and get for the amplitudes $\varepsilon_{i,0}$ a set of linear equations with a vanishing r.h.s., which has nontrivial solutions if and only if its determinant vanishes. We thus have:

$$\begin{vmatrix} 14/5 - \lambda^2 & -8/5 & -1/5 \\ -8/5 & 21/5 - \lambda^2 & -8/5 \\ -1/5 & -8/5 & 14/5 - \lambda^2 \end{vmatrix} = 0.$$

Noting $\lambda^2 = \mu/5$, we get

$$\mu^3 - 49\mu^2 + 655\mu - 2175 = (\mu - 5)(\mu - 15)(\mu - 29) = 0.$$

Indeed it is clear physically that one of the normal modes of the system is the one of the center of mass, with angular frequency ω; this corresponds to the solution $\lambda = 1$, or $\mu = 5$. We conclude that the frequencies of the normal modes read

$$\omega_1 = \omega \quad \text{(center of mass)},$$

$$\omega_2 = \sqrt{3}\,\omega,$$

$$\omega_3 = \sqrt{\frac{29}{5}}\,\omega.$$

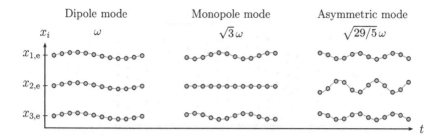

Fig. 8.3   *The three normal modes of a chain of three trapped ions, with their respective frequencies.*

In order to find the normal modes, we need to diagonalize the matrix

$$\begin{pmatrix} 14 & -8 & -1 \\ -8 & 21 & -8 \\ -1 & -8 & 14 \end{pmatrix}.$$

After a straightforward calculation, we find that the eigenvectors associated to ω, $\sqrt{3}\omega$, and $\sqrt{29/5}\omega$ are, respectively, $(1, 1, 1)$ (all the ions oscillate in phase: dipole mode), $(-1, 0, 1)$ (the two outer ions oscillate out of phase, while the central ion stands still: breathing mode) and finally $(1, -2, 1)$ (this is an asymmetric mode). Let us note that, since the matrix to be diagonalized is real and symmetric, the eigenvectors are orthogonal[9]. Figure 8.3 shows schematically the three normal modes.

An experimental example

For the cases $N = 2$ and $N = 3$, we have shown that the ratio $\omega_{\mathrm{mon}}/\omega_{\mathrm{dip}}$ is equal to $\sqrt{3}$. In the text, we are told that this results actually holds for all N (see the proof at the end of the paragraph). We thus expect, for the ratio of periods:

$$\frac{T_{\mathrm{dip}}}{T_{\mathrm{mon}}} = \sqrt{3} \simeq 1.732.$$

But Fig. 8.4 shows that we have approximately

$$\left(\frac{T_{\mathrm{dip}}}{T_{\mathrm{mon}}}\right)_{\mathrm{exp}} \simeq \frac{5}{3} \simeq 1.67,$$

---

[9]This could have helped us find the last eigenvector if we had determined the first two modes by physical arguments: existence of the dipole mode and of the symmetrical breathing mode.

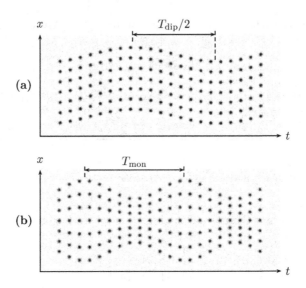

Fig. 8.4   *Experimental determination of the ratio of the periods T_{dip} of the dipole mode and T_{mon} of the monopole mode.*

i.e. only a 4% discrepancy with respect to the theoretical value. This small discrepancy is essentially due to the inaccuracy of our period measurement on the photographs.

♦ **Remark. Frequency of the breathing mode for N ions.** Let us first determine the equations fulfilled by the equilibrium positions $x_{i,\text{e}}$ ($i \in \{1, \ldots, N\}$) of the N ions. The ion i is subjected to the harmonic restoring force $-m\omega^2 x_{i,\text{e}}$, and to the repulsive Coulomb forces from the $N-1$ other ions. At equilibrium, the sum of all these forces vanishes, yielding:

$$-m\omega^2 x_{i,\text{e}} + e^2 \sum_{j=1}^{i-1} \frac{1}{(x_{i,\text{e}} - x_{j,\text{e}})^2} - e^2 \sum_{j=i+1}^{N} \frac{1}{(x_{i,\text{e}} - x_{j,\text{e}})^2} = 0, \qquad i \in \{1, \ldots, N\}.$$

We introduce the length a such that $a^3 = e^2/(m\omega^2)$, and measure all distances in units of a. The N previous equations are then rewritten as:

$$x_{i,\text{e}} = \sum_{j=1}^{i-1} \frac{1}{(x_{i,\text{e}} - x_{j,\text{e}})^2} - \sum_{j=i+1}^{N} \frac{1}{(x_{i,\text{e}} - x_{j,\text{e}})^2}, \qquad i \in \{1, \ldots, N\}. \qquad (8.8)$$

This set of algebraic equations cannot be solved in closed form for $N > 3$. However, as we shall see below, it will allow us to derive the frequency of the breathing mode for any N.

To this end, we write Newton's second law for ion i:

$$\ddot{x}_i = -x_i + \sum_{j=1}^{i-1} \frac{1}{(x_i - x_j)^2} - \sum_{j=i+1}^{N} \frac{1}{(x_i - x_j)^2}, \qquad i \in \{1, \ldots, N\}. \qquad (8.9)$$

Here, we have used a as the unit of length, and ω^{-1} as the unit of time. We focus on small oscillations around equilibrium: $x_i(t) = x_{i,\mathrm{e}} + \varepsilon(t)$ with $\varepsilon_i \ll 1$. By performing a first-order expansion we have:

$$\frac{1}{(x_i - x_j)^2} \simeq \frac{1}{(x_{i,\mathrm{e}} - x_{j,\mathrm{e}})^2}\left(1 - 2\frac{\varepsilon_i - \varepsilon_j}{x_{i,\mathrm{e}} - x_{j,\mathrm{e}}}\right).$$

Substituting into Eq. (8.9) and conserving only first-order terms, we have:

$$\ddot{\varepsilon}_i = -\varepsilon_i - 2\left(\sum_{j=1}^{i-1}\frac{\varepsilon_i - \varepsilon_j}{(x_{i,\mathrm{e}} - x_{j,\mathrm{e}})^3} - \sum_{j=i+1}^{N}\frac{\varepsilon_i - \varepsilon_j}{(x_{i,\mathrm{e}} - x_{j,\mathrm{e}})^3}\right), \qquad i \in \{1, \ldots, N\}.$$

Let us look for a solution of the form $\varepsilon_i(t) = \lambda(t)x_{i,\mathrm{e}}$, with λ *independent* of i: this corresponds to a mode for which the displacement of each ion is proportional to the distance to the center of its equilibrium position, which makes sense for a breathing mode. We then obtain:

$$\ddot{\lambda}x_{i,\mathrm{e}} = -\lambda x_{i,\mathrm{e}} - 2\lambda\left(\sum_{j=1}^{i-1}\frac{1}{(x_{i,\mathrm{e}} - x_{j,\mathrm{e}})^2} - \sum_{j=i+1}^{N}\frac{1}{(x_{i,\mathrm{e}} - x_{j,\mathrm{e}})^2}\right), \qquad i \in \{1, \ldots, N\}.$$

But, according to Eq. (8.8), the term between brackets is equal to $x_{i,\mathrm{e}}$. Simplifying by $x_{i,\mathrm{e}}$ we get the very simple equation:

$$\ddot{\lambda} = -3\lambda,$$

implying that λ oscillates in a sinusoidal way, with frequency $\sqrt{3}$. Going back to dimensional quantities, we obtain that the monopole mode frequency is $\omega\sqrt{3}$, i.e. $\sqrt{3}$ as high as that of the dipole mode[10].

Such systems of ions confined in linear Paul traps are actively studied nowadays in view of applications in the field of quantum information processing. The eigenmodes that we have just studied play a crucial role there, as they provide a coupling between the *qu-bits* (quantum bits) encoded in the internal degrees of freedom of each ion.

▶ **Further reading.** For more information, the interested reader can study the following article: D.V.F. James, *Quantum dynamics of cold trapped ions, with application to quantum computation*, Appl. Phys. B **66**, 181 (1998), also freely available on the internet at http://arxiv.org/abs/quant-ph/9702053.

---

[10]It is obvious that the dipole mode frequency is equal to the trap frequency, as is easily seen by applying Newton's second law to the system made of the N ions.

8.3 Monopole mode for interacting particles **

We consider a system made of N identical particles of mass m, confined by a two-dimensional isotropic harmonic potential:

$$U(\boldsymbol{r}) = \frac{1}{2}m\omega_0^2(x^2 + y^2).$$

We define the following *moments*:

$$M_1 = \sum_{i=1}^{N}\left(x_i^2 + y_i^2\right), \quad M_2 = \sum_{i=1}^{N}\left(x_i v_{x,i} + y_i v_{y,i}\right), \quad M_3 = \sum_{i=1}^{N}\left(v_{x,i}^2 + v_{y,i}^2\right).$$

In the absence of interactions

(1) Establish the set of linear coupled equations fulfilled by the moments M_i.

(2) Deduce the oscillation frequency of M_1 (the monopole mode, also referred to as the *breathing mode*). Comment.

In the presence of interactions

In the following, we solve the same problem but in the presence of interactions between particles. The interaction energy for a pair of particles is given by

$$V(i,j) = \frac{\alpha}{r_{ij}^2} \quad \text{with} \quad \boldsymbol{r}_{ij} = \boldsymbol{r}_i - \boldsymbol{r}_j.$$

Here, $V(i,j)$ stands for $V(\boldsymbol{r}_i, \boldsymbol{r}_j)$.

(1) Write down the conservation of the energy for the whole system.

(2) By exploiting the previous relation, find the frequency for the monopole mode from the equations of motion of the moments M_1 and M_2. Comment.

Solution

In the absence of interactions

(1) The position, velocity and acceleration of a given particle i are linked by kinematic relations and Newton's second law:

$$v_{x,i} = \frac{\mathrm{d}x_i}{\mathrm{d}t} \quad \text{and} \quad m\frac{\mathrm{d}^2 x_i}{\mathrm{d}t^2} = m\frac{\mathrm{d}v_{x,i}}{\mathrm{d}t} = -\frac{\partial U}{\partial x_i} = -m\omega_0^2 x_i.$$

From this we readily obtain the following linear set of equations:

$$\frac{dM_1}{dt} = 2M_2, \tag{8.10}$$

$$\frac{dM_2}{dt} = M_3 - \omega_0^2 M_1, \tag{8.11}$$

$$\frac{dM_3}{dt} = -2\omega_0^2 M_2. \tag{8.12}$$

For instance, let us derive explicitly the second equation:

$$\frac{dM_2}{dt} = \sum_{i=1}^{N} v_i^2 + \sum_{i=1}^{N} r_i \cdot \frac{dv_i}{dt} = \sum_{i=1}^{N} v_i^2 - \frac{1}{m} \sum_{i=1}^{N} r_i \cdot \nabla_i U(r_i),$$

where we have used Newton's second law. The first and third equations can be derived in a similar way.

(2) We search for a solution of the form $M_i \propto \exp(i\omega t)$; such a solution exists if and only if the following 3×3 determinant vanishes:

$$\text{Det} \begin{pmatrix} i\omega & -2 & 0 \\ \omega_0^2 & i\omega & -1 \\ 0 & 2\omega_0^2 & i\omega \end{pmatrix} = 4\omega_0^2 \omega - \omega^3 = 0.$$

The solutions are thus either $\omega = 0$, which simply corresponds to a static solution, and oscillations with an angular frequency $\omega = \pm 2\omega_0$. The interpretation of this second type of solutions is clear. Indeed, each component of the radius vector of a given particle x_i, y_i (and therefore each component of the velocity $v_{x,i}, v_{y,i}$) oscillates at the frequency ω_0, so that the considered moments, that are quadratic in the variables $x_i, y_i, v_{x,i}, v_{y,i}$, oscillate at $2\omega_0$. Combining the differential equations (8.10), (8.11) and (8.12), one easily finds the equation

$$\frac{d^3 M_1}{dt^3} + 4\omega_0^2 \frac{dM_1}{dt} = 0,$$

whose solution reads $M_1 = M_0 + \delta M \sin(2\omega_0 t)$ (after an appropriate choice of the origin of time). In the case of small amplitude oscillations $\delta M \ll M_0$, the evolution of the mean radius $R = M_1^{1/2}/N$ is therefore given by

$$R = R_0 + \frac{\delta M}{2NR_0} \sin(2\omega_0 t).$$

One readily understands why this mode is referred to as the "breathing mode" since it corresponds to an oscillation of the mean radius of the cloud around a given value R_0.

In the presence of interactions

(1) The conservation of the total energy of the N interacting particles reads

$$E_0 = \sum_{i=1}^{N} \left(\frac{1}{2}mv_i^2 + \frac{1}{2}m\omega_0^2 r_i^2 \right) + \frac{\alpha}{2}\sum_{i\neq j} \frac{1}{r_{ij}^2}.$$

The factor $1/2$ in the interaction term avoids the double counting of particle pairs.

(2) The first relation (8.10) is unchanged in the presence of interactions. The equation for the moment M_2 reads

$$\frac{dM_2}{dt} = \sum_{i=1}^{N} v_i^2 - \frac{1}{m}\sum_{i=1}^{N} r_i \cdot \nabla_i \left(U(r_i) + \sum_{j\neq i} V(i,j) \right)$$

$$\frac{dM_2}{dt} = \frac{2}{m}\sum_{i=1}^{N} [E_c(i) - U(r_i)] - \frac{1}{m}\sum_{i=1}^{N} r_i \cdot \nabla_i \left(\sum_{j\neq i} V(i,j) \right),$$

where $E_c(i) = mv_i^2/2$. Let us denote by J the last term of the previous equation. To calculate this term we proceed step by step. First, we notice that:

$$\nabla_i V(i,j) = \frac{\partial}{\partial r_i} V(i,j) = -2\alpha \frac{r_i - r_j}{r_{ij}^4}.$$

Combining this relation with the definition of J, one obtains:

$$J = \frac{2\alpha}{m} \sum_{i=1}^{N}\sum_{j\neq i} r_i \cdot \frac{r_i - r_j}{r_{ij}^4}.$$

The summation over the two indices i and j is symmetric, so that the sum is unchanged by a permutation of the indices:

$$J = \frac{2\alpha}{m}\frac{1}{2} \sum_{i=1}^{N}\sum_{j\neq i} \left(r_i \cdot \frac{r_i - r_j}{r_{ij}^4} + r_j \cdot \frac{r_j - r_i}{r_{ij}^4} \right)$$

$$J = \frac{\alpha}{m} \sum_{i=1}^{N}\sum_{j\neq i} \frac{1}{r_{ij}^2}$$

$$J = \frac{2}{m}\left(E_0 - \sum_i (E_c(i) + U(r_i)) \right).$$

The equation of motion for M_2 can therefore be put in the form

$$\frac{dM_2}{dt} = \frac{1}{2}\frac{d^2 M_1}{dt^2} = \frac{2E_0}{m} - \frac{4}{m}\sum_i U(r_i).$$

Using Eq. (8.10), we finally obtain

$$\frac{\mathrm{d}^2 M_1}{\mathrm{d}t^2} + 4\omega_0^2 M_1 = \frac{4E_0}{m}.$$

We thus recover an oscillation at $2\omega_0$, as in the case without interactions! This is a special feature[11] of a two-body interaction potential scaling as r^{-2}. For other power-law potentials, we would not have obtained this result. This is one of the very few examples in which one can extract in a simple way an exact result about a N-body system in the presence of interactions. Another example (with the same interaction potential, which is not a coincidence) is given by the Calogero–Sutherland model (see Problem 6.3).

8.4 Mass spectrometry **

Mass spectrometry is a technique allowing one to sort ions according to their charge-to-mass ratio. If the ions are created from a sample of an unknown material, one can determine its composition quantitatively.

A mass spectrometer is composed of three main parts located inside a vacuum chamber: an ion source, which creates ions to be analyzed starting from the sample introduced into the spectrometer, a *mass analyzer*, which selects the ions depending on their charge-to-mass ratio, and finally an ion detector, connected to a recorder. The main goal of this exercise is to study the working principle of three different mass analyzing devices, namely the *time-of-flight analyzer*, the *magnetic sector*, and finally the *quadrupole analyzer*.

Production and detection of the ions

(1) A widely used technique for the production of the ions to be analyzed, starting from atoms or molecules, consists in bombarding them with electrons produced by a high temperature filament and accelerated by an electric field. What is the order of magnitude of the energy that one needs to communicate to these electrons?

(2) How can one detect a flux of ions?

---

[11] L.P. Pitaevskii and A. Rosch, *Breathing modes and hidden symmetry of trapped atoms in two dimensions*, Phys. Rev. A **55**, R853 (1997), available at http://arxiv.org/abs/cond-mat/9608135.

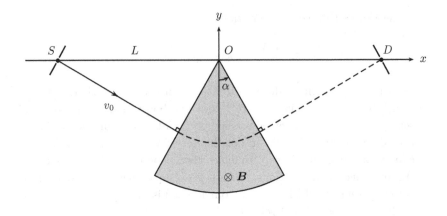

Fig. 8.5   *Magnetic sector mass spectrometer.*

Mass analyzers

In what follows we consider positively charged ions with mass m and charge ze (e is the elementary charge, $z \in \mathbb{N}$ is the ionisation degree). We suppose, for the sake of simplicity, that the initial velocity of the ions is negligible after ionisation.

Time of flight.

The time-of-flight analyzer is composed of an acceleration zone made by a parallel plate capacitor where the ions are accelerated by an electrostatic potential V_0. The detector is located at a distance d from the output of the acceleration zone. The source is switched on for a very short time interval and we measure the distribution of the arrival times τ of the ions. Show that this allows to infer the distribution of the ratios ze/m of the ions created in the source.

Magnetic sector.

After being accelerated by a potential V_0, the ion beam, collimated by a source slit S (Fig. 8.5), goes through an angular sector of opening angle 2α in which a uniform magnetic field B deflects the particles. An ion detector is placed behind a slit D, symmetric of S with respect to the magnetic sector. The ion beam reaches the magnetic sector at normal incidence.

(1) Show that in this way one detects selectively the ions having a given

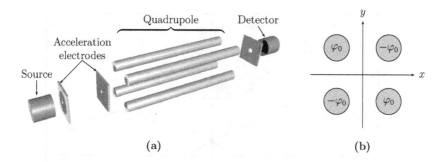

Fig. 8.6 *Quadrupole mass spectrometer.* (a) *General view.* (b) *Transverse cut of the quadrupole.*

ratio ze/m.

(2) How can one change the type of ions detected, in order to obtain a full mass spectrum?

(3) The ion beam has necessarily some divergence at the output of the source slit S. Show that the angular geometry allows to get a focusing effect. For that purpose, consider a trajectory coming from S and making an angle $\varepsilon \ll 1$ with respect to the normal, and show that, to first order in ε, the ions hit the slit D.

(4) The accelerating voltage in this type of spectrometer is on the order of $V_0 \simeq 2$ kV. What is the required field B in order to have a trajectory of radius 50 cm, for an ion with a mass 100 amu[12], with $z = 1$? Comment.

Quadrupole analyzer.

In the quadrupole analyzer (Fig. 8.6), invented by Wolfgang Paul in the fifties, the ions accelerated along the z-axis are subjected to the electric field produced by four parallel electrodes (on which potentials $\pm\varphi_0$ are applied). Close to the axis, the electrostatic potential reads:

$$\varphi(x, y) = \varphi_0 \frac{x^2 - y^2}{R^2},$$

where R is a characteristic length. One uses for φ_0 a DC voltage U onto which is superimposed an AC component oscillating at a high frequency Ω:

$$\varphi_0(t) = U + V \cos(\Omega t).$$

---

[12]We recall that the atomic mass unit, or amu, is defined as one twelfth of the mass of a ^{12}C atom, or about 1.66×10^{-27} kg. It is close to (actually slightly smaller than) the mass of a nucleon (proton or neutron).

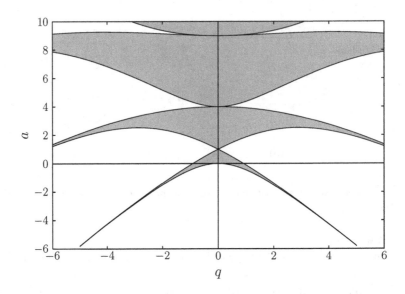

Fig. 8.7   *Stability diagram of Mathieu's equation (8.13) in the (q, a) plane. The shaded areas correspond to stable solutions.*

(1) Explain *qualitatively* why the DC component U acts as "low-pass" mass filter for the ions: those having a high m/ze ratio are accelerated towards the electrodes and thus lost, and why the AC component of amplitude V acts, on the contrary, as a "high-pass" mass filter.

(2) In order to perform a more quantitative analysis, write down the equations of motion for the transverse coordinates (x, y) of the ion. By introducing adequate dimensionless parameters, rewrite these equations as the standard Mathieu equation:

$$\frac{d^2\xi}{d\tau^2} + [a - 2q\cos(2\tau)]\,\xi = 0\,. \tag{8.13}$$

(3) Mathieu's differential equation is linear and of second order, but has non-constant coefficients; its solutions cannot be expressed in terms of elementary functions. However, one can show that, as a function of the parameters (q, a), the solutions are either oscillating (and thus bounded), either diverging. The former are said to be stable, and the latter unstable. Figure 8.7 shows, in the plane (q, a), the various stability regions. Comment on this diagram.

(4) Show graphically that for fixed (U, V, Ω), the ions have a stable transverse motion, both along x and along y, provided one is in precise

regions of the (q, a) plane. In this diagram, where are ions having different ratios ze/m located? Show that one can thus select ions according to their charge-to-mass ratio, with a very good selectivity. How can one vary the value of the ze/m ratio selected by the quadrupole analyzer?

(5) What are the advantages of such an analyzer as compared to the magnetic sector?

Solution

Production and detection of the ions

(1) During the electronic bombardment, the atoms or molecules to be analyzed must become ions. Therefore, the energy of the incident electrons must exceed the ionization energy E_I of the atom or molecule; this energy typically ranges from a few eV to a few tens of eV (it is well known for example that $E_I \simeq 13.6$ eV for the hydrogen atom). In practice, one uses electrons with a kinetic energy on the order of 100 eV.

(2) In order to detect a flux of ions, one collects them onto a metallic electrode which is kept, by an external circuit, to a constant potential. The arrival of charges on the electrode then creates a current in this circuit. Measuring this current gives a direct determination of the flux of ions. In practice, one sometimes uses, especially in the case of low fluxes, electron multipliers similar to the ones used in photomultipliers.

Mass analyzers

Time of flight.

If we neglect the initial velocity of the ions after their creation from the atoms or molecules to be analyzed[13], their final velocity v_f fulfills (by energy conservation):

$$\frac{1}{2}mv_f^2 = zeV_0 \qquad \Longleftrightarrow \qquad v_f = \sqrt{\frac{2zeV_0}{m}}.$$

---

[13]Which implies that the acceleration voltages are large enough, for instance on the order of a kilovolt, so that one can neglect the initial kinetic energy, i.e. a few ionization energies, compared to the final kinetic energy.

The time of flight τ required to move over a distance d then reads:

$$\tau = \frac{d}{v_{\mathrm{f}}} = \sqrt{\frac{m}{ze}}\, \frac{d}{\sqrt{2V_0}}.$$

It is clear that the time of flight depends (as the square root) of the charge-to-mass ratio ze/m of the ion. By measuring the distribution $p(\tau)$ of the arrival times of the ions, one can easily infer the distribution $f(m/ze)$ of their charge-to-mass ratios.

Let us notice however that for realistic sizes ($d \sim 1$ m), and for a singly-ionized ion with a mass $m = 1$ amu, accelerated by $V_0 \sim 1$ kV, the time of flight is typically $\tau \sim 2$ μs. This implies to use a pulsed source, that can be switched on and off over much shorter times. This technique thus requires relatively fast electronics.

Magnetic sector.

(1) In the magnetic sector, the ions have a circular trajectory, whose radius ρ_{L} (the so-called Larmor radius) reads:

$$\rho_{\mathrm{L}} = \frac{mv_{\mathrm{f}}}{zeB} = \sqrt{\frac{m}{ze}}\, \frac{\sqrt{2V_0}}{B}.$$

Given the geometry of the device, only the ions having a trajectory whose radius of curvature fulfills $\rho_{\mathrm{L}} = L\sin\alpha$ reach the detector D; therefore, only the ions with a charge-to-mass ratio

$$\frac{ze}{m} = \frac{2V_0}{B^2 L^2 \sin^2\alpha}$$

are detected.

(2) If the geometry of the device is fixed (which is obviously the most convenient experimentally), we can vary the ratio ze/m of the selected ions by modifying either the acceleration voltage V_0, either the magnetic field B.

(3) Because of the divergence of the ionic beam, some ions do not arrive at normal incidence (at I) onto the magnetic sector, but make a small angle $\varepsilon \ll 1$ with respect to normal incidence; they reach the magnetic sector at point I' (Fig. 8.8).

We will show that, to first order in ε, the ion trajectory does reach the detector D. For that purpose, we shall show that the center of the circular trajectory of the ion in the sector is, to first order in ε, located on the axis y. In this case, the full trajectory of the ion has Oy

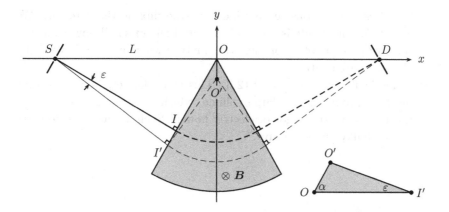

Fig. 8.8 *Magnetic sector mass spectrometer: focusing effect. On the right, the triangle $OO'I'$ is shown with a very exaggerated angle ε.*

as a symmetry axis, which implies that the ion reaches D (the point symmetric of S with respect to Oy).

Let O' be the intersection of the line perpendicular to SI' going through I' and of the y-axis. The center of the circular trajectory of the ion belongs to[14] $(O'I')$. In order to show that, to first order in ε, the center of the trajectory is O', it is then sufficient for us to prove that $O'I' = \rho_L + o(\varepsilon)$. In the following, we will do all the calculations to first order in ε. Obviously, $OI' = \rho_L + L\varepsilon\cos\alpha = \rho_L(1 + \varepsilon/\tan\alpha)$. In the triangle $OO'I'$, by writing the "law of sines":

$$\frac{O'I'}{\sin\widehat{O'OI'}} = \frac{OI'}{\sin\widehat{OO'I'}},$$

we get $O'I' = OI'\sin\alpha/\sin(\pi - \alpha - \varepsilon) \simeq OI'(1 - \varepsilon/\tan\alpha)$ to first order. Whence:

$$O'I' = \rho_L(1 + \varepsilon/\tan\alpha)(1 - \varepsilon/\tan\alpha) = \rho_L + o(\varepsilon),$$

which ends our proof. Note that, of course, instead of using the geometric arguments presented here, one can study analytically the trajectories of the ions, by solving their equations of motion, and recover the same result.

The geometry of the magnetic sector has thus a major advantage: it allows for a *focusing* effect (analogous to the one of a converging lens

---

[14]Since the velocity of the ion is continuous at the input of the magnetic sector (although its derivative is not).

in optics) which considerably increases the flux on the detector. We note that this result is true only in the limit of small angles (in the same way as, in optics, one loses stigmatism when one goes beyond the Gauss approximation).

(4) Numerically, we find $B \simeq 0.12$ T. This is a relatively high field, that must be produced in a large volume, which thus requires an electro-magnet dissipating a lot of electric power. Such a mass spectrometer is thus bulky and expensive.

Quadrupole analyzer.

(1) Let us consider an ion with a large mass, which thus has a large in-ertia. The characteristic timescale for its motion in a given electric field is therefore also large. This means that the AC voltage V, which varies rapidly on this timescale, has a very weak influence on the ion motion, for its effects almost average out to zero. We thus neglect this component; only the DC component U then contributes to the motion of the ion. Now, the field corresponding to the electrostatic potential $\varphi \propto (x^2 - y^2)$ is confining along one direction but repulsive along the other. The ion will thus be attracted towards a pair of electrodes and will not reach the detector. The quadrupole thus behaves, because of the DC component U, as a low-pass mass filter.

We now consider on the contrary an ion with a very weak mass. It then reacts very quickly to the variations of the electric potential applied on the electrodes. We can thus "forget" about the time depen-dence of the potential, and consider that the motion of the ion takes place in the *instantaneous* quadrupole potential. But we have just seen above that this motion is unstable along one or the other of the direc-tions x and y; the ion will thus be lost. Therefore the quadrupole does not let the very light ions reach the detector either; it is thus also a high-pass mass filter.

Finally, we can thus expect qualitatively that the quadrupole an-alyzer is a band-pass mass filter, which selects the ions with a mass m such that the characteristic timescale of the motion in a field φ_0/R, namely $\sqrt{mR^2/(ze\varphi_0)}$, is comparable to Ω^{-1}. The goal of the following questions is to prove this property.

(2) Newton's second law, projected onto the axes x and y, yields:

$$m\ddot{x} = zeE_x = -ze\frac{\partial\varphi}{\partial x}$$

$$m\ddot{y} = zeE_y = -ze\frac{\partial\varphi}{\partial y}\,,$$

or, by writing explicitly the time and position dependence of φ,

$$\ddot{x} = -\frac{2ze}{mR^2}\left(U + V\cos\Omega t\right)x \tag{8.14}$$

$$\ddot{y} = \frac{2ze}{mR^2}\left(U + V\cos\Omega t\right)y\,. \tag{8.15}$$

We consider first Eq. (8.14). Noting

$$\tau \equiv \frac{\Omega t}{2}\,, \quad a \equiv \frac{8zeU}{m\Omega^2R^2}\,, \quad \text{and} \quad q \equiv \frac{-4zeV}{m\Omega^2R^2}\,, \tag{8.16}$$

we get

$$x'' + (a - 2q\cos 2\tau)x = 0\,, \tag{8.17}$$

where the independent variable is now τ: this is exactly Mathieu's equation. For the equation of motion (8.15) along y, we get

$$y'' - (a - 2q\cos 2\tau)y = 0\,, \tag{8.18}$$

which is also a Mathieu's equation, of parameters $(-q, -a)$.

(3) We first notice that for $q = 0$, Mathieu's equation reduces to

$$x'' + ax = 0\,,$$

in which we recognize, for $a > 0$, the equation of a harmonic oscillator of angular frequency $\sqrt{a}$. It thus makes sense that the half-line ($q = 0, a > 0$) corresponds to stable trajectories. In the same way, for ($q = 0, a < 0$), the solutions diverge exponentially; this half-line thus corresponds to unstable trajectories.

For all q, whatever small, there are some instability regions around $a = n^2$ ($n \in \mathbb{N}^*$). They correspond to the phenomenon called *parametric resonance*. The lowest one, for $a = 1$, corresponds to a periodic modification of the eigenfrequency of the oscillator, at a frequency which is twice the eigenfrequency: this resonance is exactly the one that the reader was exploiting during his/her childhood, in order to increase the oscillation amplitude on a swing!

Let us notice also that there exists a stable region (with a relatively small area, however) for $a < 0$. The modulation can thus stabilize a motion which otherwise would be unstable (see also Problems 4.1 and 8.6); this is the working principle of the quadrupole analyzer.

Finally, we can notice that for $a = 0$, the motion is stable up to a critical value $q \simeq 0.9$. Let us show that the motion is indeed stable in the limit $q \ll 1$. This would correspond to a slow motion (with angular frequency $\sqrt{q}$) in a potential where the cos term would be replaced by 1. We thus look for a solution of the form $\xi = \xi_{\text{slow}} + \xi_{\text{fast}}$ to the equation of motion

$$\ddot{\xi} = 2q \cos(2\tau)\xi \,,$$

where ξ_{fast} evolves with a frequency on the order of one; while ξ_{slow} evolves with a much lower frequency. By conserving only the "fast" terms, we get

$$\ddot{\xi}_{\text{fast}} = 2q \cos(2\tau)\xi_{\text{slow}} \,,$$

which is readily integrated as (ξ_{slow} is considered as constant at this stage):

$$\xi_{\text{fast}} = -\frac{q}{2} \cos(2\tau)\xi_{\text{slow}} \,.$$

Substituting this result into the equation of motion and keeping only the slow terms (and also using the fact that the average value of $\cos^2(2\tau)$ is simply $1/2$), we obtain:

$$\ddot{\xi}_{\text{slow}} = -\frac{q^2}{2}\xi_{\text{slow}} \,,$$

an equation showing that the slow motion is harmonic, with an angular frequency $q/\sqrt{2} \ll 1$ (this legitimates *a posteriori* the separation of the motion into two components, a slow one and a fast one).

(4) For an ion to reach the detector, its transverse motion (along x and y) has to be stable. From Eqs (8.17) and (8.18), the parameters (q, a) defined in Eq. (8.16) must therefore be such that (q, a) and $(-q, -a)$ correspond to stable solutions. Graphically, this corresponds, in the plane (q, a), to the intersection of the stable zones of the stability diagram of Fig. 8.7 and of its symmetric with respect to the origin. The largest of these intersection zones is magnified on Fig. 8.9.

We shall show that this transverse stability criterion allows one to select the ions depending on their ratio ze/m. Indeed, let us assume that the parameters U, V and Ω characterizing the voltages applied on the electrodes are fixed. It is then obvious that q and a are proportional to ze/m, the proportionality constant being fixed by U, V and Ω. Ions

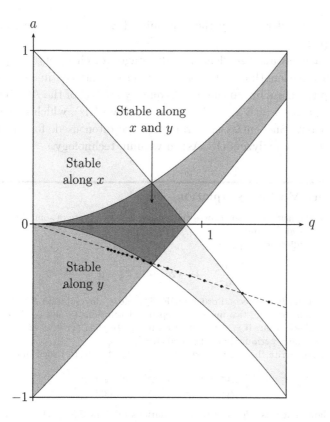

Fig. 8.9 *The stability domain (dark gray) in the (q, a) plane for a quadrupole mass filter is the intersection of the two zones corresponding to a stable motion along x (light gray) and along y (gray). The points (q_m, a_m) corresponding to ions of mass m are aligned onto a line (whose slope $-V/(2U)$ is determined by the voltages U and V applied to the electrodes), and their abscissa is proportional to 1/m (with a proportionality factor varying as $1/\Omega^2$). In this way, by changing Ω, one can sweep the mass of the detected ions. The ratio U/V determines the width of the intersection between this line and the stability zone, and thus allows one to chose the selectivity of the mass analyzer (its ability to separate ions with almost equal masses). In practice, one thus chooses the slope in a way such that the line crosses the stability zone in the immediate neighborhood of the lower angle of the stability zone.*

differing by their ze/m ratio are thus aligned, in the plane (q, a), onto a straight line going through the origin (see Fig. 8.9). By choosing the ratio U/V one fixes the slope of this line; one chooses it in such a way that its intersection with the stability zone is narrow enough so that a single charge-to-mass ratio is transmitted. One thus realizes a mass filter. In order to vary the transmitted ze/m value, one can simply sweep Ω.

(5) Such a mass analyzer has the advantage of being very compact (the quadrupole length is on the order of ten to twenty centimeters); one can sweep the mass by varying the frequency $\Omega/(2\pi)$ of the AC component (a frequency which is on the order of a few MHz), which is technically very easy. Such mass spectrometers are often used, for instance, as residual gas analyzers (RGAs) in vacuum technology.

Reminder: Mathieu's equation

Here, we recall quickly how one proves the existence of two types of solutions, stable and unstable, to Mathieu's equation.

Mathieu's differential equation

$$\frac{d^2\xi}{d\tau^2} + [a - 2q\cos(2\tau)]\,\xi = 0 \qquad (8.19)$$

is a homogeneous *linear* second-order ODE. Therefore, any solution ξ can be written as a linear combination of two linearly independent solutions ξ_1 and ξ_2. Moreover, one immediately checks that, if $\xi(\tau)$ is a solution to (8.19), then $\xi(\tau + \pi)$ is also a solution (this is due to the π-periodicity of the coefficients).

We can thus write that there exist constants (A_1, A_2, B_1, B_2) such that:

$$\begin{cases} \xi_1(\tau + \pi) = A_1\xi_1(\tau) + A_2\xi_2(\tau), \\ \xi_2(\tau + \pi) = B_1\xi_1(\tau) + B_2\xi_2(\tau). \end{cases} \qquad (8.20)$$

We shall show that this allows us to find solutions fulfilling $\Xi(\tau + \pi) = \mu\Xi(\tau)$ where μ is a (complex) constant. By writing $\Xi = \alpha\xi_1 + \beta\xi_2$ and substituting into Eq. (8.20), we obtain an homogeneous system of two equations for α and β:

$$\begin{cases} (A_1 - \mu)\alpha + B_1\beta = 0, \\ A_2\alpha + (B_2 - \mu)\beta = 0. \end{cases}$$

We have non-trivial solutions if and only if $(A_1 - \mu)(B_2 - \mu) - B_1 A_2 = 0$. This quadratic equation for μ has in general two complex solutions μ_1 and μ_2. Let us note Ξ_i $(i = 1, 2)$ the solution corresponding to μ_i. One easily sees that the solution to the functional equation

$$\Xi_i(\tau + \pi) = \mu_i\Xi_i(\tau) \qquad (8.21)$$

reads

$$\Xi_i(\tau) = \mu_i^{\tau/\pi}\chi_i(\tau),$$

where χ_i is any π-periodic function[15].

---

[15]To see that, we define $\chi_i(\tau) \equiv \mu_i^{-\tau/\pi}\Xi_i(\tau)$, and we calculate $\chi_i(\tau + \pi)$:

$$\chi_i(\tau + \pi) = \mu_i^{-\tau/\pi}\mu_i^{-1}\Xi_i(\tau + \pi) = \mu_i^{-\tau/\pi}\mu_i^{-1}\mu_i\Xi_i(\tau) = \chi_i(\tau),$$

which does show that χ_i is an arbitrary π-periodic function.

We shall now establish two properties of the coefficients μ_i. On the one hand, we have $\mu_1\mu_2 = 1$. This results from the following property of the *Wronskian*[16] $W(\tau)$ of Ξ_1 and Ξ_2:

$$W(\tau) = \text{const.} = W(\tau + \pi),$$

which is easily obtained from Mathieu's equation. But, using Eq. (8.21), we have $W(\tau + \pi) = \mu_1\mu_2 W(\tau)$, which implies[17] that $\mu_1\mu_2 = 1$. On the other hand, the coefficients of Mathieu's equation being real, we can conclude that the set $\{\mu_1, \mu_2\}$ coincides with $\{\mu_1^*, \mu_2^*\}$.

We thus have two possibilities:

(1) One has $\mu_1 = \mu_1^*$ and $\mu_2 = \mu_2^*$, they are thus real; from $\mu_1\mu_2 = 1$ we deduce $\mu_1 = e^{\sigma\tau}$ and $\mu_2 = e^{-\sigma\tau}$ with σ real. The solutions have the form $e^{\sigma\tau}\chi_1(\tau)$ and $e^{-\sigma\tau}\chi_2(\tau)$, which shows that we have an *unstable* solution.

(2) We have $\mu_1 = \mu_2^*$; from $\mu_1\mu_2 = 1$ we get $|\mu_1| = |\mu_2| = 1$, thus $\mu_1 = e^{i\sigma\tau}$ and $\mu_2 = e^{-i\sigma\tau}$ where σ is real. The solutions are $e^{i\sigma\tau}\chi_1(\tau)$ and $e^{-i\sigma\tau}\chi_2(\tau)$, showing one has *stable* solutions.

▶ **Further reading.** The article written by Wolfgang Paul, *Electromagnetic traps for charged and neutral particles*, Rev. Mod. Phys. **62**, 531 (1990), after he was awarded the Nobel Prize in Physics in 1989 is a good introduction to the physics of quadrupole mass spectrometers. The reader may also consult W. Paul's Nobel lecture, freely available at http://nobelprize.org/nobel_prizes/physics/laureates/1989/paul-lecture.html.

8.5 Ponderomotive force **

The aim of this exercise is to study the role played by the transverse intensity profile of a laser beam on the motion of a charged particle. The maximum intensity is assumed to be sufficiently low so that charged particles are not accelerated by the laser field to velocities close to the speed of light. The motion will therefore be described by non-relativistic mechanics. Under this assumption, we simply have to take into account the action of the electric field of the laser on the charged particle. We thus consider the motion of a particle of charge q and mass m in the following oscillating electric field: $\boldsymbol{E} = \boldsymbol{E}_0(r)\sin(\omega t)$.

---

[16]The Wronskian W of two functions Ξ_1 and Ξ_2 is defined as $W = \dot{\Xi}_1\Xi_2 - \dot{\Xi}_2\Xi_1$. If it vanishes, the two functions are linearly dependent (for their logarithmic derivatives are then equal).

[17]The Wronskian is nonzero, since we assumed that Ξ_1 and Ξ_2 are linearly independent.

(1) Write down the vectorial equations of motion of the charged particle.

(2) To extract the underlying physics, we search for a solution of this equation in the form: $r(t) = r_0(t) + \delta r(t)$, where δr evolves on the fast timescale $T = 2\pi/\omega$ of this problem, and r_0 is supposed to evolve on a much longer timescale. We assume also that $|\delta r| \ll |r_0|$.

   (a) Show that the first-order expansion of the electric field reads:

$$E_0(r) = E_0(r_0) + (\delta r \cdot \nabla)E_0(r_0).$$

   (b) Deduce the differential equation for the time evolution of $\delta r(t)$ at first order. Solve this equation with the following initial conditions: $\delta \dot{r}(0) = -qE_0(r_0)/(m\omega)$ and $\delta r(0) = 0$.

   (c) Show that the motion of the slow component $r_0(t)$ can be described as the one of a particle evolving in a static potential. We recall the vectorial calculus identity

$$(a \cdot \nabla)\, a = \nabla \left(a^2/2\right) - a \times (\nabla \times a). \tag{8.22}$$

   (d) The intensity of the laser beam is proportional to the square of the electric field $E_0(r)$. This intensity is maximal in the center of the beam, and decreases radially. Describe qualitatively the motion of a charged particle that experiences a laser pulse. Does the motion depend on the sign of the charge?

(3) An intense laser pulse is focused onto a gas of neutral helium atoms. Explain qualitatively why the laser pulse tends to defocus[18]. We recall the expression for the refractive index of a plasma:

$$n(r, z, t) = \left(1 - \frac{\omega_p^2}{\omega^2}\right)^{1/2} = \left(1 - \frac{n_e(r, z, t)}{n_c}\right)^{1/2},$$

where n_e is the electronic density and n_c a critical density, here equal to 10^{21} cm^{-3}. The angular frequency ω_p is referred to as the *plasma frequency* and was already mentioned in Problem 1.2.

Solution

(1) Newton's second law reads:

$$m\ddot{r} = qE_0(r)\sin(\omega t).$$

[18]T. Auguste, P. Monot, L.-A. Lompré, G. Mainfré, and C. Manus, *Defocusing effects of a picosecond tera-watt laser pulse in an underdense plasma*, Opt. Commun. **89**, 145 (1992).

(2) (a) In order to derive the desired vectorial relation, let us consider the x component. The first-order expansion yields

$$E_{0x}(x_0 + \delta x, y_0 + \delta y, z_0 + \delta z) \simeq$$
$$E_{0x}(x_0, y_0, z_0) + \left(\delta x \frac{\partial}{\partial x} + \delta x \frac{\partial}{\partial y} + \delta z \frac{\partial}{\partial z} \right) E_{0x}(x_0, y_0, z_0).$$

The generalization to the two other components is straightforward.

(b) Combining this relation with the equation of motion, we get

$$m\ddot{\boldsymbol{r}}_0 + m\delta\ddot{\boldsymbol{r}} = [q\boldsymbol{E}_0(\boldsymbol{r}_0) + q(\delta\boldsymbol{r} \cdot \boldsymbol{\nabla})\boldsymbol{E}_0] \sin(\omega t). \tag{8.23}$$

According to the text, the term $\delta\boldsymbol{r}$ oscillates at the angular frequency ω. In Eq. (8.23), the first term of the right hand side oscillates at ω; the second term has a time dependence that originates from the product of $\delta r(t)$ by $\sin(\omega t)$. This latter term gives rise to[19] a term that oscillates at 2ω, plus a constant term. Identifying in Eq. (8.23) the terms that oscillate at the same angular frequency ω we find:

$$m\delta\ddot{\boldsymbol{r}} = q\boldsymbol{E}_0(\boldsymbol{r}_0) \sin(\omega t).$$

Using the initial conditions, we have:

$$\delta r(t) = -\frac{q}{m\omega^2} \boldsymbol{E}_0(\boldsymbol{r}_0) \sin(\omega t). \tag{8.24}$$

(c) Averaging over one optical period $T = 2\pi/\omega$ relation (8.23) removes all the terms that oscillate at frequencies ω and 2ω. It thus remains:

$$m\ddot{\boldsymbol{r}}_0 = \langle q(\delta\boldsymbol{r} \cdot \boldsymbol{\nabla})\boldsymbol{E}_0 \sin(\omega t) \rangle .$$

Using the explicit form of $\delta\boldsymbol{r}$ from Eq. (8.24), we eventually obtain

$$\ddot{\boldsymbol{r}}_0 = -\frac{q^2}{m^2\omega^2} (\boldsymbol{E}_0 \cdot \boldsymbol{\nabla})\boldsymbol{E}_0 \langle \sin^2(\omega t) \rangle.$$

The mean value over one period of $\langle \sin^2(\omega t) \rangle$ is:

$$\langle \sin^2(\omega t) \rangle = \frac{\omega}{2\pi} \int_0^{2\pi/\omega} \sin^2(\omega t) dt = \frac{1}{2}.$$

Using Eq. (8.22), we can rewrite the term $(\boldsymbol{E}_0 \cdot \boldsymbol{\nabla})\boldsymbol{E}_0$ as

$$(\boldsymbol{E}_0 \cdot \boldsymbol{\nabla})\boldsymbol{E}_0 = \boldsymbol{\nabla} \left(\frac{E_0^2}{2} \right) + \boldsymbol{E} \times \frac{\partial \boldsymbol{B}}{\partial t}$$

---

[19] Indeed, one has $\sin(\omega t)\sin(\omega t) = [1 - \cos(2\omega t)]/2$ and $\sin(\omega t)\cos(\omega t) = \sin(2\omega t)/2$.

(where we have used Maxwell-Faraday equation $\nabla \times \boldsymbol{E} = -\partial_t \boldsymbol{B}$). In the non-relativistic regime, the effect of the magnetic field $\boldsymbol{B}$ on the motion is negligible as compared to the one of the electric field $\boldsymbol{E}$, so that one has:

$$(\boldsymbol{E}_0 \cdot \nabla)\boldsymbol{E}_0 \simeq \nabla \left(\frac{E_0^2}{2} \right).$$

In summary, the low-frequency component of the motion obeys the following effective equation of motion:

$$m\ddot{\boldsymbol{r}}_0 = -\nabla \left(\frac{q^2 E_0^2}{4m\omega^2} \right) = -\nabla[V(r_0)].$$

This equation is the one of a particle that experiences a potential $V(r) = q^2 E_0^2 / 4m\omega^2$. The resulting force depends on the spatial inhomogeneities of the electric field, and thus of the intensity profile of the laser. For a plane wave, this force vanishes.

(d) Using the form of the potential that we have derived, we can predict that the charged particles will be expelled from the center of the beam. Indeed, transversally, the center of the beam appears as the top of the potential hill experienced by the charged particle. This position is therefore unstable. Note that this effect does not depend on the sign of the charge, since it is the square of the charge that appears in the formula. The larger the mass, the longer the timescale for expelling the particles. This is a simple inertial effect. The motion of a charged particle in the field of a laser therefore contains two terms: a slow motion, dictated by the intensity profile, and a rapid motion with a small amplitude at the optical frequency.

(3) For an intense laser pulse propagating in a gas, the above discussion makes sense only if the rising edge of the pulse ionizes the particles of the (initially neutral) gas. The ionization threshold for helium is about 10^{14} W·cm^{-2} for a laser with a wavelength of 1064 nm (a multi-photonic absorption process is responsible for the ionization). As a result, the electron density builds up faster on the laser beam axis than on the edge, which leads to a local decrease of the index of refraction. This inhomogeneous plasma acts as a diverging lens that yields a defocusing of the pulse. This effect is illustrated in Fig. 8.10. In the plane π_1, the atoms constituting the gas are not ionized yet, while in the plane π_2, they are, and the defocusing process takes place.

A spectacular effect on the optical propagation of the pulse, referred to as the relativistic self-channeling effect, occurs when the plasma

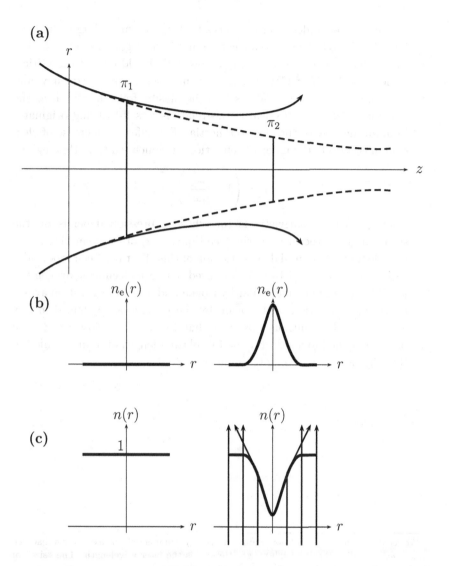

Fig. 8.10 (a) *Focusing of the laser pulse in the vacuum (dashed line) and in the gas ionized by the front of the pulse (solid line). As a result of the local modification of the refractive index the beam is defocused.* (b) *Transverse profile of the electronic density $n_e(r)$ at positions π_1 and π_2.* (c) *Refractive index $n(r)$ at positions π_1 and π_2.*

is sufficiently dense and is made of electron having relativistic velocities. In this regime, and contrary to the effect studied in this exercise, the pulse modifies locally the index of refraction in such a way that it can be guided over distances that can be much larger than the Rayleigh length[20] one would get from classical gaussian optics. For this to happen, the incident laser power P should exceed the critical value $P_c = 17\omega^2/\omega_p^2$ GW, where ω is the laser angular frequency, and $\omega_p = 5.64 \times 10^4 (n_e/\text{cm}^{-3})^{1/2}$ s^{-1} is the plasma frequency, with n_e the electronic density. Physically, the relativistic self-focusing originates from an increase of the electron inertia. This effective increase of electron mass affects the index of refraction, n, which reads, in this regime,

$$
n = \left(1 - \frac{\omega_p^2}{\gamma\omega^2} \right)^{1/2} ,
$$

where γ is the relativistic factor whose expression depends on the strength of the vector potential (see Eq. (11.22) of Problem 11.4). The first clear experimental demonstration of this effect has been reported[21] with a 400 fs pulse of 15 TW[22] focused onto a molecular hydrogen gas jet. The use of both molecular hydrogen and a jet configuration avoids premature refraction of the beam by the plasma, as explained in this exercise[23]. The authors observed that for a power close to P_c, the beam remained focused over the full plasma length of 3 mm, while the Rayleigh length was only 0.3 mm!

---

[20]We recall that a Gaussian laser beam of waist w_0 remains well focused over a distance $z_R = \pi w_0^2/\lambda$, called the Rayleigh range. Here, λ is the laser wavelength. The "size" of the beam changes by only a factor $\sqrt{2}$ over this distance.

[21]P. Monot, T. Auguste, P. Gibbon, F. Jakober, G. Mainfray, A. Dulieu, M. Louis-Jacquet, G. Malka, and J.L. Miquel, *Experimental demonstration of relativistic self-channeling of a multiterawatt laser pulse in an underdense plasma*, Phys. Rev. Lett. **74**, 2953 (1995).

[22]These conditions correspond to a maximum intensity, measured in vacuum, of about 3×10^{18} W · cm^{-2}.

[23]The defocusing effect drastically reduces the peak intensity from its value in the vacuum and limits the achievable plasma density. As a result, it is not possible to achieve a sufficiently high plasma density for the available power.

8.6 Effective potential and phase jump **

The goal of this problem is to introduce the notion of *effective potential* for the slow motion of a particle moving in a rapidly oscillating time-dependent potential $U(x,t) = U_0(x)\cos(\omega t + \varphi)$, and to show that it is possible to decrease exponentially the energy of a particle mass m, by changing the phase φ at well-chosen times. The latter idea was recently proposed theoretically[24], but not yet demonstrated experimentally.

We will restrict ourselves to a one-dimensional problem. The potential is obtained *via* a periodic modulation of the factor $U_0(x) = kx^2/2$, and we shall assume that $\omega \gg (k/m)^{1/2}$.

(1) Write down the equation of motion.
(2) The variable $x(t)$ is decomposed into two components $x(t) = X(t) + \xi(t)$ where $X(t)$ varies slowly on a timescale ω^{-1}, and $\xi(t)$ has a frequency ω and has (as its derivatives) a vanishing time average over a period $2\pi/\omega$:

$$\langle \xi \rangle = \frac{\omega}{2\pi} \int_0^{2\pi/\omega} \xi(t')\, \mathrm{d}t' = 0.$$

Calculate $\mathrm{d}^2\xi/\mathrm{d}t^2$, $\mathrm{d}\xi/\mathrm{d}t$ and $\xi(t)$ as a function of k, m, ω and $X(t)$.
(3) Show that the dynamics of the slow variable X reduces to that of a particle moving in the time-independent effective potential $U_{\text{eff}}(X)$:

$$U_{\text{eff}}(X) = \frac{k^2 X^2}{4m\omega^2}.$$

(4) Calculate the average kinetic energy of the *micro-motion* associated with the fast variable ξ, as a function of $U_{\text{eff}}(X)$. Comment.
(5) One decides at time τ to change abruptly the phase from $\varphi(t = \tau^-) = 0$ to $\varphi(t = \tau^+) = \pi/2$. We assume, without loss of generality, that $\omega\tau$ is a multiple of 2π. Write down the energy variation ΔE for the slow variable resulting from the phase jump, as a function of m and of the velocities $\mathrm{d}X/\mathrm{d}t(\tau^\pm) \equiv \dot{X}(\tau^\pm)$.
(6) Show that it is possible, by choosing in a clever way the time at which the phase jump occurs, to reduce significantly the energy E of the slow motion.
(7) Deduce from the above results an optimal method to decrease the energy of a particle in this context. Comment.

---

[24] A. Ridinger and N. Davidson, *Particle motion in rapidly oscillating potentials: the role of the potential's initial phase*, Phys. Rev. A **76**, 013421 (2007); also freely available on the internet at the address http://arxiv.org/abs/0705.2737.

It is *a priori* surprising that a potential of the form $U(x,t) = U_0(x)\cos(\omega t + \varphi)$ allows for trapping a particle. Indeed the potential $U(x,t)$ switches between attractive and repulsive depending on the sign of $\cos(\omega t + \varphi)$, and, over a period $2\pi/\omega$, has vanishing average: $\langle U(x,t)\rangle = 0$. However, the result mentioned in the text is valid when $\omega \gg (k/m)^{1/2}$, i.e. in the limit where the rate at which the sign changes is large as compared to the oscillation frequency of the particle in the potential well $U(x)$. The motion of the particle can be decomposed in that case as the sum of a fast, low amplitude component ξ, oscillating at a frequency ω, and of a much slower component (whose characteristic timescale will be determined below).

(1) The equation of motion is obtained from Newton's second law:

$$m\frac{d^2x}{dt^2} = -\frac{d}{dx}U(x,t) = -kx\cos(\omega t + \varphi). \qquad (8.25)$$

(2) We substitute $x(t) = X(t) + \xi(t)$ into Eq. (8.25):

$$\frac{d^2X}{dt^2} + \frac{d^2\xi}{dt^2} = -\frac{k}{m}(X + \xi)\cos(\omega t + \varphi). \qquad (8.26)$$

If we now average over a period $2\pi/\omega$, Eq. (8.26) becomes:

$$\frac{d^2\langle X\rangle}{dt^2} = \frac{d^2X}{dt^2} = -\frac{k}{m}\langle \xi\cos(\omega t + \varphi)\rangle. \qquad (8.27)$$

Indeed, over a period $2\pi/\omega$ the slow variable $X(t)$ stays almost constant, and thus $\langle X\rangle(t) \simeq X(t)$. The same is true for the derivatives of X. The remaining term $\langle \xi\cos(\omega t + \varphi)\rangle$ has a low frequency contribution, i.e. on long timescales, as it is the product of two fast variables and can thus have a non-vanishing average, since $\langle \cos^2(\omega t + \varphi)\rangle = 1/2$. Subtracting Eq. (8.27) from Eq. (8.26), we get:

$$\frac{d^2\xi}{dt^2} = -\frac{k}{m}X\cos(\omega t + \varphi) - \frac{k}{m}\left[\xi\cos(\omega t + \varphi) - \langle \xi\cos(\omega t + \varphi)\rangle\right].$$

The last term has only frequency components at 2ω. Indeed, assuming that $\xi \sim \cos(\omega t + \varphi)$, we find

$$\xi\cos(\omega t + \varphi) - \langle \xi\cos(\omega t + \varphi)\rangle \sim \cos^2(\omega t + \varphi) - \langle \cos^2(\omega t + \varphi)\rangle$$

$$= \frac{1}{2}\left[\cos(2\omega t + 2\varphi) + 1\right] - \frac{1}{2} = \frac{\cos(2\omega t + 2\varphi)}{2}.$$

We shall neglect these contributions that introduce harmonics at $n\omega$ ($n > 1$). Note that these harmonics have their contribution attenuated when one integrates twice $d^2\xi/dt^2$ with respect to time in order to get $\xi(t)$, since their amplitude is then divided by n^2.

We finally get a very simple equation for the fast variable ξ:

$$\frac{d^2\xi}{dt^2} \simeq -\frac{k}{m}X\cos(\omega t + \varphi).$$

Integrating this equation is trivial as the separation of timescales allows us to consider that X stays almost constant over an oscillation period $2\pi/\omega$:

$$\frac{d\xi}{dt} = -\frac{k}{m\omega}X\sin(\omega t + \varphi) \quad \text{and} \quad \xi = \frac{k}{m\omega^2}X\cos(\omega t + \varphi). \tag{8.28}$$

The constants of integration were taken equal to zero since the time averages of ξ and $d\xi/dt$ are assumed to be zero.

(3) The explicit expression of ξ can now be substituted into Eq. (8.27):

$$m\frac{d^2X}{dt^2} = -k\langle \xi\cos(\omega t + \varphi)\rangle = -k\left\langle \frac{k}{m\omega^2}X\cos^2(\omega t + \varphi)\right\rangle,$$

$$m\frac{d^2X}{dt^2} = -\frac{k^2}{m\omega^2}X\langle\cos^2(\omega t + \varphi)\rangle = -\frac{k^2}{2m\omega^2}X,$$

$$m\frac{d^2X}{dt^2} = -\frac{d}{dX}\left(\frac{k^2X^2}{4m\omega^2}\right) = -\frac{d}{dX}U_{\text{eff}}(X). \tag{8.29}$$

We thus show that the dynamics of the slow variable X reduces to that of a particle moving in the effective time-independent harmonic potential $U_{\text{eff}}(X)$. This potential is sometimes called the *ponderomotive* potential. Two other problems of the present book, "Mass spectrometry" (Problem 8.4, see the section about quadrupolar mass spectrometers), and "Ponderomotive force" (Problem 8.5), emphasize the interest of this concept. The characteristic timescale in this potential is $2\pi/\Omega$ with

$$U_{\text{eff}}(X) = \frac{k^2X^2}{4m\omega^2} = \frac{1}{2}m\Omega^2X^2.$$

Our hypothesis of separated timescales thus amounts to:

$$\Omega \ll \omega, \quad \text{or} \quad \omega \gg \sqrt{\frac{k}{m}}.$$

(4) It is interesting at this point to evaluate the energy stored in the fast motion. We thus calculate the average kinetic energy for the fast variable ξ:

$$\left\langle \frac{1}{2}m\left(\frac{d\xi}{dt}\right)^2\right\rangle = \frac{1}{2}m\left(\frac{k}{m\omega}X\right)^2\langle\sin^2(\omega t + \varphi)\rangle = U_{\text{eff}}(X).$$

The amplitude of the fast motion stays small, but due to its high frequency, the velocity is high, resulting in an average kinetic energy equal to the potential energy $U_{\text{eff}}(X)$ of the (large amplitude) slow motion.

(5) One decides at time τ to change abruptly the phase from $\varphi(t = \tau^-) = \varphi^- = 0$ to $\varphi(t = \tau^+) = \varphi^+ = \pi/2$. We are interested in the slow variable X. The amplitude of ξ is small since $\omega \gg (k/m)^{1/2}$. As the phase φ affects the variable ξ, we can make the approximation: $X(t = \tau^-) = X(t = \tau^+)$. However, we have seen that the kinetic energy of the micro-motion is on the same order of magnitude as the characteristic energy of the slow variable. We thus cannot make the same approximation for the velocity:

$$\left(\frac{dX}{dt}\right)_{t=\tau^-} = \dot{X}(\tau^-) = \dot{x}(\tau^-) - \dot{\xi}(\tau^-, \varphi^-)$$

$$\left(\frac{dX}{dt}\right)_{t=\tau^+} = \dot{X}(\tau^+) = \dot{x}(\tau^+) - \dot{\xi}(\tau^+, \varphi^+)$$

The true velocity is continuous $\dot{x}(\tau^-) = \dot{x}(\tau^+)$, as, from Eq. (8.25), the acceleration undergoes a *finite* discontinuity during the phase jump $\varphi^- \longrightarrow \varphi^+$. We conclude that:

$$\dot{X}(\tau^+) = \dot{X}(\tau^-) + \dot{\xi}(\tau^-, \varphi^-) - \dot{\xi}(\tau^+, \varphi^+). \tag{8.30}$$

The energy E associated with the slow variable is, according to Eq. (8.29):

$$E = \frac{1}{2}m\dot{X}^2 + U_{\text{eff}}(X). \tag{8.31}$$

The total variation ΔE of the energy of the slow variable thus reads:

$$\Delta E = E(\tau^+) - E(\tau^-) = \frac{1}{2}m[\dot{X}^2(\tau^+) - \dot{X}^2(\tau^-)].$$

(6) From Eqs (8.28) and (8.30), the maximal difference in velocities for a well chosen time τ is:

$$\dot{X}(\tau^+) = \dot{X}(\tau^-) + \frac{kX(\tau)}{m\omega},$$

as the cosine varies from 1 to 0. The corresponding variation of energy is

$$\Delta E = \frac{1}{2}m[\dot{X}(\tau^+) - \dot{X}(\tau^-)][\dot{X}(\tau^+) - \dot{X}(\tau^-) + 2\dot{X}(\tau^-)]$$

$$\Delta E = \frac{1}{2}m\frac{kX(\tau)}{m\omega}\left(\frac{kX(\tau)}{m\omega} + 2\dot{X}(\tau^-)\right). \tag{8.32}$$

From energy conservation (8.31), applied at time τ^-, the velocity $\dot{X}(\tau^-)$ reads:

$$\dot{X}(\tau^-) = \pm\sqrt{\frac{2E}{m} - \frac{k^2 X^2}{2m^2\omega^2}}. \tag{8.33}$$

We notice that the sign cannot be determined *a priori*, whence the $\pm$ sign.

It is convenient to use at this stage dimensionless quantities in order to gain more physical insight. It is natural here to introduce the reduced variable $x = X/X_0$ where X_0 is the amplitude of the slow motion before the phase jump: $E = U_{\text{eff}}(X_0)$, and thus $X_0^2 = 4m\omega^2 E/k^2$. By substituting the solution (8.33) into (8.32), we obtain the simple result:

$$\Delta E_\pm(x) = 2E\left(x^2 \pm x\sqrt{2}\sqrt{1-x^2}\right).$$

The energy variation ΔE depends on the sign of the velocity $\dot{X}(\tau^-)$ of the slow variable before the phase jump, and on the amplitude $x = X(\tau)/X_0$ at the time when the phase jump occurs.

We look for the extrema of ΔE_+:

$$\frac{\mathrm{d}}{\mathrm{d}x}\Delta E_+(x) = 0 = 2E\left(2x - \frac{\sqrt{2}x^2}{\sqrt{1-x^2}} + \sqrt{2}\sqrt{1-x^2}\right).$$

We thus need to solve the equation:

$$x\sqrt{2}\sqrt{1-x^2} + 1 - 2x^2 = 0. \tag{8.34}$$

Introducing the auxiliary variable $u = 2x^2 - 1$, Eq. (8.34) becomes:

$$\sqrt{1-u^2} = \pm u\sqrt{2} \qquad \Longrightarrow \qquad u = \pm\frac{1}{\sqrt{3}}.$$

We deduce that the solutions fulfill:

$$x_\pm^2 = \frac{1}{2}\left(1 \pm \frac{1}{\sqrt{3}}\right).$$

Substituting into Eq. (8.34), we see that two of the above solutions have to be discarded, and we get two solutions:

$$x_+ = \sqrt{\frac{1}{2}\left(1 + \frac{1}{\sqrt{3}}\right)} \quad \text{and} \quad x_- = -\sqrt{\frac{1}{2}\left(1 - \frac{1}{\sqrt{3}}\right)}.$$

The corresponding energy variation reads:

$$\frac{\Delta E_+(x_+)}{E} \simeq 2.73 \qquad \text{and} \qquad \frac{\Delta E_+(x_-)}{E} \simeq -0.73.$$

The same result is obtained for $\Delta E_-(-x_\pm)/E$. Finally, a single phase jump allows us to make the energy of the particle decrease significantly:

$$\frac{E(\tau^+)}{E(\tau^-)} \simeq 0.27.$$

Note that we can also "pump" energy into the particle motion, since $\Delta E_+(x_+) > E$.

(7) We simply need to iterate the procedure. After n phase jumps at appropriate times, the final energy $E_f(n)$ is only $E_f(n) = 0.27^n E_i$ where E_i is the initial energy associated with the slow variable. Note that this exponential decrease of E can at best be performed in a time which is on the order of one oscillation period. Indeed, just after a phase jump, the position of the particle does not have the right value for the next jump. By letting it evolve for a time smaller than the period, it reaches the position where the jump leads to an optimum decrease in energy, and so on. However, implementing this method requires a perfect knowledge of the position of the particle: this is obviously an obstacle to the experimental realization. Note also that this is the reason why the second law of thermodynamics is not violated in this experiment: the experimenter, who acts as the famous Maxwell's demon, needs to extract information about the position of the particle in order to reduce its energy.

Chapter 9

"Cold" Atoms

Since the beginning of the eighties, considerable progress has been made in the manipulation of the external degrees of freedom (position and velocity) of neutral atoms. These techniques of laser cooling and trapping have revolutionized atomic physics, as they allow us to replace thermal atomic beams or glass cells containing a vapor at room temperature, with atomic clouds containing from a few million to a few billion atoms, at temperatures below a few microkelvin! Many useful applications have emerged; let us cite only one here: the most accurate atomic clocks (with relative accuracies better than 10^{-15}) now use these techniques.

The progress in this field of physics was recognized by the award of the Nobel prize in Physics in 1997 to Claude Cohen-Tannoudji, Steven Chu, and William D. Phillips, for their pioneering work in laser cooling[1]. More recently, in 1995, a combination of laser cooling and of another technique (called evaporative cooling) has allowed for the observation of *Bose–Einstein condensation*, a purely quantum phenomenon predicted in 1924 by Albert Einstein. The Nobel prize in Physics 2001 was awarded to Eric A. Cornell, Carl E. Wieman, and Wolfgang Ketterle for the achievement of Bose–Einstein condensation in dilute gases[2]. In the following problems, one studies simple models of some of the techniques allowing us to obtain what is called — improperly[3] — "cold" atoms.

---

[1] See http://nobelprize.org/nobel_prizes/physics/laureates/1997/.

[2] See http://nobelprize.org/nobel_prizes/physics/laureates/2001/.

[3] Strictly speaking, it is the atomic vapor, and not the atoms themselves, that is cold.

9.1 Effusive atomic beam **

We consider a chamber (called "oven" in the following), at temperature T, with a small aperture of area A_1. It contains a metal in equilibrium with its vapor at the saturated vapor pressure $P_{sat}(T)$ corresponding to the temperature T of the oven; we will note n the corresponding atomic density (number of atoms per unit volume). The atoms of the vapor can escape through the aperture into a vacuum chamber containing the oven (see Fig. 9.1).

We assume that the dimensions of the aperture (on the order of $\sqrt{A_1}$) are small as compared to the mean free path of the atoms inside the oven. This implies that the velocity distribution of the atoms in the vapor is not changed by the presence of the aperture, it is thus the equilibrium, Maxwell-Boltzmann distribution. This regime, called *effusive*, is valid as long as the vapor density is not too high (i.e. at relatively low temperature).

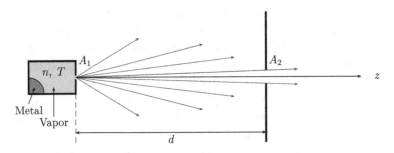

Fig. 9.1 *Atomic beam in the effusive regime, produced using an oven at temperature T.*

(1) (a) Calculate the flux Φ of the atomic beam thus realized (i.e. the number of atoms leaving the oven per unit time).

  (b) What is the corresponding numerical value for an oven containing rubidium, with an aperture of area $A_1 = 1$ mm^2, at a temperature of 100 °C (we give the rubidium saturation pressure at this temperature: 2.7×10^{-2} Pa, as well as the mass of a rubidium atom: $m = 1.45 \times 10^{-25}$ kg).

  (c) How long can one use the oven, assuming that, initially, a mass $m_0 = 1$ g of rubidium was introduced into it?

(2) (a) What is the angular distribution of the atomic beam obtained in this way?

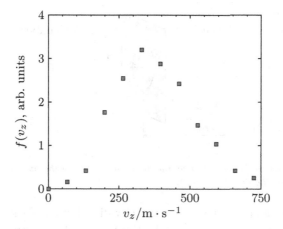

Fig. 9.2 *Longitudinal velocity distribution of a rubidium effusive beam, measured by laser spectroscopy.*

(b) In order to have a more directional beam, one *collimates* the beam by means of a second aperture, with area A_2, located at a distance d from the first one (see Fig. 9.1). What is the flux Φ_c of the collimated beam?

(c) What is the numerical value of Φ_c for $d = 10$ cm and $A_2 = 25$ mm^2? What is the order of magnitude of the beam divergence?

(3) (a) Calculate the longitudinal velocity distribution $f(v_z)$ of the beam. What is the most probable longitudinal velocity?

(b) Figure 9.2 shows the velocity distribution for an effusive beam of rubidium atoms. It has been obtained by sweeping the frequency of a counterpropagating laser beam, resonant with an atomic transition, and by measuring the intensity of the fluorescence light emitted by the atoms (because of the Doppler effect, the resonance frequency depends on the atomic velocity, which allows us to reconstruct $f(v_z)$). Deduce the temperature T of the oven.

Solution

(1) (a) In order to calculate the atomic flux, we will calculate the number of atoms leaving the oven per unit time *for a given velocity* (magnitude and direction), and then integrate over velocities. In

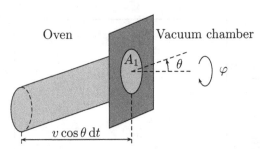

Fig. 9.3 *Volume swept during* dt *by the atoms going out of the oven with a velocity* **v** *oriented towards the direction* (θ, φ).

cartesian coordinates, the velocity distribution of the atoms in the vapor in the oven is given by the Maxwell-Boltzmann distribution:

$$f(v_x, v_y, v_z) = \mathcal{N} \exp\left(-\frac{m(v_x^2 + v_y^2 + v_z^2)}{2k_B T}\right),$$

where the constant $\mathcal{N}$ is obtained by the normalization condition:

$$\int f(v_x, v_y, v_z)\, d^3v = 1 \quad \Longrightarrow \quad \mathcal{N} = \left(\frac{m}{2\pi k_B T}\right)^{3/2}.$$

In this calculation we have used the well-known result concerning the value of the Gauss integral[4]: $\int_{-\infty}^{\infty} \exp(-x^2)\, dx = \sqrt{\pi}$.

It is convenient to use spherical coordinates, with, as the polar axis, the normal to the oven aperture (see Fig. 9.3). The probability of having a velocity of magnitude v (within dv) and heading towards the spherical angles (θ, φ) (within the elementary solid angle $d^2\Omega = \sin\theta\, d\theta\, d\varphi$) then reads

$$d^3P = \mathcal{N} v^2 \exp\left(-\frac{mv^2}{2k_B T}\right) dv\, d^2\Omega,$$

since the volume element in velocity space is $dv_x\, dv_y\, dv_z = v^2 \sin\theta\, dv\, d\theta\, d\varphi$. The number d^4N of atoms going out of the oven during dt, with a velocity having a magnitude v (within dv), and directed towards (θ, φ) (within $d\theta$ and $d\varphi$) is thus given by the product of the number of atoms being in the oblique cylinder shown on Fig. 9.3 (this number is $nA_1 v\, dt\, \cos\theta$) by the probability d^3P of having the right velocity. We thus find:

$$d^4N = nA_1 v \cos\theta\, d^3P\, dt. \tag{9.1}$$

---

[4]See the reminder at the end of the solution.

To obtain the flux, we just have to integrate over φ (between 0 and 2π), over θ (between 0 and $\pi/2$ — not π, see Fig. 9.3!) and over the magnitude of the velocity (between 0 and ∞). The number $dN = \Phi\,dt$ of atoms leaving the oven during dt, independently of their velocity, is thus:

$$dN = nA_1\,dt\,\mathcal{N}2\pi \int_0^{\pi/2} \sin\theta\cos\theta\,d\theta \int_0^\infty v^3\exp\left(-\frac{mv^2}{2k_BT}\right)dv.$$

The angular integral equals $1/2$. The integral over v can be calculated by posing $x = mv^2/(2k_BT)$ and integrating by parts. We finally get:

$$\Phi = \frac{1}{4}nA_1\bar{v}$$

where $\bar{v}$ is the mean velocity of the atoms *in the vapor*:

$$\bar{v} = \frac{\displaystyle\int_0^\infty v^3\exp\left(-\frac{mv^2}{2k_BT}\right)dv}{\displaystyle\int_0^\infty v^2\exp\left(-\frac{mv^2}{2k_BT}\right)dv} = \sqrt{\frac{8k_BT}{\pi m}}.$$

(b) Numerically, we obtain for the mean velocity $\bar{v} \simeq 300$ m $\cdot$ s^{-1}, and for the vapor density (assuming that one has an ideal gas, which is obviously a very good approximation for such a low pressure) $n = P_{\text{sat}}(T)/(k_BT) \simeq 5.2\times10^{18}$ m^{-3}; this yields a flux $\Phi \simeq 4\times10^{14}$ s^{-1}.

(c) As long as there is a drop of liquid rubidium remaining in the oven (the melting point of rubidium is 33 $^\circ$C), the density n stays constant, and therefore so does the flux. The oven "lifetime" is therefore

$$\tau = \frac{N_0}{\Phi} = \frac{m_0/m}{\Phi} \simeq 2\times10^7\,\text{s},$$

or about 200 days.

♦ **Remarks.**

(i) When only some vapor is remaining inside the oven, the decrease in the number of remaining atoms N is exponential (for in that case the density, and thus the flux, is proportional to N: we get the differential equation $\Phi = \dot{N} = N/\tau'$). The time constant is $\tau' = 4V/(A_1\bar{v})$, with V the volume of the oven. For $V = 1$ L, we get $\tau' \simeq 13$ s. The oven thus "stops" very quickly when the last droplet of liquid rubidium has disappeared.

(ii) Let us check that for this oven temperature, the beam is indeed in the effusive regime. The mean free path is $\ell_{\text{mfp}} = 1/(n\sigma)$ where σ is the collision cross section. For a monoatomic gas at room temperature we have $\sigma \sim 10^{-18}$ m^2. Therefore $\ell_{\text{lpm}} \sim 20$ cm, which is very large compared to the size of the oven aperture.

(2) (a) It clearly appears in Eq. (9.1) that the angular distribution of the beam follows a law in $\cos\theta$. Although the beam intensity is maximal on axis (the normal to the oven aperture), we thus have a very diverging beam, which is therefore difficult to use.

(b) In the approximation where the distance of the collimator fulfills $d^2 \gg A_2$, we simply need to calculate the flux of atoms corresponding to a velocity directed towards the small solid angle $\delta\Omega = A_2/d^2$ around the direction $\theta = 0$. We find immediately:

$$\Phi_{\rm c} = \frac{nA_1 A_2 \bar{v}}{4\pi d^2}.$$

(c) Numerically, as $\Phi_{\rm c} = \Phi A_2/(\pi d^2)$, we get $\Phi_{\rm c} \simeq 3 \times 10^{11}$ s^{-1}. The divergence of the beam is on the order of $\sqrt{A_2}/d \sim 5/100$ (or about $3°$). The beam thus has a relatively well-defined propagation direction. This is obtained, obviously, at the expense of the flux, which is reduced here by more than three orders of magnitude.

(3) (a) The distribution of longitudinal velocities can be read directly in Eq. (9.1), for $\theta \simeq 0$. We have:

$$f(v_z) \propto v_z^3 \exp\left(-\frac{mv_z^2}{2k_{\rm B}T}\right). \tag{9.2}$$

In a beam, the atoms of high velocity are more probable than in a vapor (the velocity distribution in the vapor simply reads $v^2 \exp(-mv^2/(2k_{\rm B}T))$, and thus is less "peaked" towards high velocities), because a larger number of fast atoms leave the oven in a given time interval (indeed, the cylinder of Fig. 9.3 has a volume proportional to v). The most probable velocity v_0 is the one for which the function $f(v_z)$ as a maximum. We find easily:

$$v_0 = \sqrt{\frac{3k_{\rm B}T}{m}}. \tag{9.3}$$

♦ **Remark.** The most probable velocity is not the only quantity characterizing the velocity distribution of the atoms: one can calculate also for instance the mean velocity and the rms velocity. For the effusive beam, they read respectively[5] $\sqrt{9\pi k_{\rm B}T/(8m)}$ and $\sqrt{4k_{\rm B}T/m}$.

(b) On the experimental data, we observe that the most probable velocity is $v_0 \simeq 350$ m $\cdot$ s^{-1}. Equation (9.3) then gives for the oven temperature $T \simeq 400$ K. In order to confirm this estimate in a more accurate way, Fig. 9.4 shows the result of a fit of the data by a function of the form (9.2), with the temperature as the only

---

[5] The reader is invited to perform the calculation by him/herself.

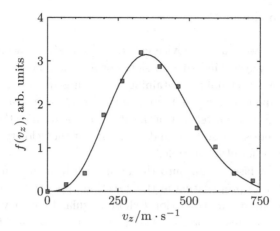

Fig. 9.4 *A fit of the experimental data by the theoretically expected distribution; the temperature obtained in this way is* 410 K.

adjustable parameter. The agreement with experimental points is excellent; we obtain a temperature $T \simeq 410$ K in good agreement with the estimate above.

Reminder: Gaussian integral

We want to calculate the value of the integral

$$I = \int_{-\infty}^{\infty} \exp(-x^2)\, dx.$$

For that, we calculate the square

$$I^2 = \left(\int_{-\infty}^{\infty} \exp(-x^2)\, dx \right) \times \left(\int_{-\infty}^{\infty} \exp(-y^2)\, dy \right)$$

$$= \iint_{\mathbb{R}^2} \exp\left[-(x^2 + y^2) \right]\, dx\, dy$$

where the last equality is obtained using Fubini's theorem. Now, the latter integral is easy to calculate in polar coordinates:

$$I^2 = 2\pi \int_{0}^{\infty} \exp(-r^2)\, r\, dr = \pi,$$

which proves the result given in the text.

9.2   Zeeman slower **

In this exercise, we study a device called *Zeeman slower*, which is widely used in atomic physics in order to slow down atomic beams. By drilling an aperture into a chamber containing a low pressure vapor and letting the atoms escape into a vacuum, one obtains an *effusive atomic beam* (see Problem 9.1). The mean velocity of the atoms in the beam is then of several hundreds of meters per second, and the width of the velocity distribution is of the same order of magnitude.

By shining a laser beam onto the atoms (with angular frequency ω_L^0 and wavevector $\mathbf{k}$) propagating in the direction opposite to the beam and resonant with an atomic transition (whose angular frequency is ω_A^0), one exerts onto the atoms a force, called the *radiation pressure* force, which allows one to decelerate them. One can thus almost put the beam to rest, as well as decrease its velocity dispersion. In this way, one produces very slow atoms that can be used for experiments that would be impossible to perform with fast atoms. In the following, we model the deceleration induced by the radiation pressure using a one-dimensional model, along the axis (Ox) of the beam.

(1) **A few properties of the radiation pressure force.**

    (a) We assume that the atom absorbs γ photons per second from the laser. By using momentum conservation, evaluate the force exerted onto the atom.

    (b) For an atomic transition whose excited state has a lifetime $1/\Gamma$, the maximum photon scattering rate is $\gamma = \Gamma/2$. What is the numerical value of the maximum acceleration undergone by a ^{87}Rb atom (for the transition used, we have $\Gamma/(2\pi) \simeq 6$ MHz and the laser wavelength is $\lambda = 780$ nm).

    (c) Assuming that the radiation pressure force always keeps its maximum value, what is the length ℓ needed to bring to rest an atom with initial velocity $v_0 = 500$ m · s^{-1}? How many photons are scattered during deceleration?

    (d) In fact, the radiation pressure force is maximum only when the frequency ω_L of the laser corresponds to the resonance frequency ω_A of the atomic transition. When the *detuning* $\delta = \omega_L - \omega_A$ is nonzero, the force exerted onto the atom reads:

$$F = F_0 \frac{1}{1 + 4\delta^2/\tilde{\Gamma}^2},$$

where $\tilde{\Gamma}$ is on the same order of magnitude as Γ (actually, slightly larger than Γ). Plot a graph of F/F_0 as a function of δ. What is its full width (at half-maximum)?

(2) **Slowing down a resonant particle.**

(a) We consider a particle with initial velocity v_i, initially in resonance with the laser, and we assume that the detuning is always zero (without bothering for now about how this condition is realized). Write down the velocity $v(x)$ of the atom as a function of its position.

(b) Because of the Doppler effect, the laser frequency "seen" by the atom depends on its velocity v: $\omega_L = \omega_L^0 + kv$. By how much can the atom velocity change without altering significantly the value of the radiation pressure force? Calculate numerically this change in velocity.

(c) In order to be able to decelerate the atom all the way to zero velocity, one uses a spatially varying magnetic field $B(x)$. Due to the Zeeman effect, the atomic resonance frequency reads $\omega_A = \omega_A^0 + \mu B(x)/\hbar$. What is the functional form of $B(x)$ such that the resonance condition is always fulfilled?

(d) Calculate numerically the variation ΔB in magnetic field between the input and the output of the slower, assuming that the variation of velocity is $500\,\mathrm{m \cdot s^{-1}}$ and that, for the atomic transition, $\mu = \mu_B$ (μ_B is the Bohr magneton, which has the numerical value $\mu_B = h \times 1.4 \times 10^{10}\,\mathrm{Hz \cdot T^{-1}}$).

(3) **Cooling effect.** In practice, when one designs a Zeeman slower, one chooses a value a for the acceleration (slightly smaller than the value a_0 corresponding to zero detuning[6]), the initial velocity v_i of the atoms to decelerate, and the final velocity v_f of those atoms at the output of the slower (this value is dictated by the experiment to be done with the beam).

(a) Show that these conditions fix the length ℓ of the slower.

(b) Write down the magnetic field profile $B(x)$ as a function de a, v_i, v_f, k, μ, x and ℓ.

(c) This magnetic field profile $B(x)$ depends explicitly on v_i. But the atomic beam has, at the output of the slower, a very broad velocity distribution. What happens to a particle with an initial velocity

---

[6]This allows for imperfections in the magnetic field profile.

smaller or larger than v_i? One may use a graphical representation in phase space (x, v).

(d) Plot qualitatively the velocity distribution of the beam obtained at the output of the slower, in the case where v_i is on the same order of magnitude as the mean velocity of the initial beam. Comment. Does the name "Zeeman *slower*" capture all the aspects of the operation of such a device?

Solution

(1) **A few properties of the radiation pressure force.**

(a) When an atom absorbs a photon, whose linear momentum is $\hbar \boldsymbol{k}$ (see reminder below), its own momentum varies by the same quantity. If the atom absorbs γ photons per unit time, the time derivative of its momentum is thus

$$\frac{\mathrm{d}\boldsymbol{p}}{\mathrm{d}t} = \gamma \hbar \boldsymbol{k}.$$

But, from Newton's second law, the force exerted on the atom is nothing but $\mathrm{d}\boldsymbol{p}/\mathrm{d}t$. The radiation pressure force thus reads:

$$\boldsymbol{F}_{\mathrm{rad}} = \gamma \hbar \boldsymbol{k}.$$

♦ **Remark.** Actually, after every absorption of a photon, the atom re-emits a photon by spontaneous emission into a random direction; its momentum also varies due to this spontaneous emission. However, the probabilities of emission into two opposite directions are equal; and thus, over a large number of cycles of absorption-spontaneous emission, the recoil effect due to spontaneous emission averages to zero. On the contrary, the direction of the momentum of the absorbed photons is always the same (it is given by the direction of the laser beam) and thus the recoil effect is cumulative.

(b) Since the maximum photon scattering rate is $\gamma = \Gamma/2$, the maximum acceleration for an atom of mass m reads

$$a_{\mathrm{max}} = \frac{F_{\mathrm{max}}}{m} = \frac{\hbar k \Gamma}{2m} = \frac{\pi \hbar \Gamma}{m\lambda} \simeq 10^5 \, \mathrm{m \cdot s^{-2}}.$$

This is a huge acceleration (10^4 times the acceleration due to gravity!) which allows one to slow down, over a very short distance, atoms that are initially very fast, as we shall see below.

(c) For a uniformly decelerated motion with acceleration $-a_{\mathrm{max}}$, the velocity depends on the distance x as

$$v(x) = \sqrt{v_0^2 - 2a_{\mathrm{max}}x}.$$

The length ℓ that we seek corresponds to $v(\ell) = 0$; therefore we have:

$$\ell = \frac{v_0^2}{2a_{\max}} \simeq 1.25\,\mathrm{m}.$$

In order to calculate the number N of scattered photons, we find the time $\tau = v_0/a_{\max}$ needed for deceleration, and we multiply it by the scattering rate $\Gamma/2$:

$$N = \frac{v_0\Gamma}{2a_{\max}} = \frac{v_0}{v_{\mathrm{rec}}}$$

where, in the last equality we introduced the *recoil velocity* $v_{\mathrm{rec}} = \hbar k/m \simeq 6\,\mathrm{mm}\cdot\mathrm{s}^{-1}$, which is nothing but the decrease in velocity due to the absorption of a single photon. We finally find $N \sim 10^5$ photons.

(d) The behavior of the radiation pressure force

$$F = F_0 \frac{1}{1 + 4\delta^2/\tilde{\Gamma}^2}.$$

as a function of the detuning δ is shown in Fig. 9.5. This a *Lorentzian* curve, with a full width at half-maximum $\tilde{\Gamma}$.

♦ **Remark.** The width of this resonance curve is not equal to Γ because of the so-called *radiative broadening*: when the intensity I of the laser is large (as compared to a quantity called *saturation intensity*, noted I_{sat} and whose order of magnitude is about one milliwatt per square centimeter), the resonance width becomes $\tilde{\Gamma} = \Gamma\sqrt{1 + I/I_{\mathrm{sat}}}$. In practice, in usual Zeeman slowers, one rarely goes beyond $\tilde{\Gamma} \sim 4\Gamma$ or so, as the available laser power is obviously limited.

(2) **Slowing down a resonant particle.**

(a) We deal with a uniformly accelerated motion (actually, decelerated); the velocity thus fulfills:

$$v(x) = \sqrt{v_{\mathrm{i}}^2 - 2a_0 x}, \tag{9.4}$$

where $a_0 = F_0/m$.

(b) The radiation pressure force becomes negligible when the detuning between the atomic resonance and the frequency of the laser beam becomes on the order of a few $\tilde{\Gamma}$, which, because of the Doppler effect, corresponds to a velocity change of a few

$$\Delta v_{\mathrm{Doppler}} = \frac{\tilde{\Gamma}}{k} \gtrsim \frac{\Gamma\lambda}{2\pi} \simeq 5\,\mathrm{m}\cdot\mathrm{s}^{-1}.$$

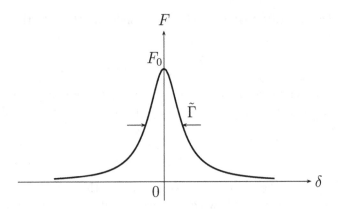

Fig. 9.5   *Variation of the radiation pressure force as a function of the detuning δ.*

This value is very small compared to the typical initial velocities of the atom (several hundreds of meters per second). The atom therefore decelerates under the influence of the radiation pressure, but, very quickly, its velocity has decreased too much for it to still be resonant, and the deceleration stops.

(c) In order to obtain for all values of x the resonance condition $\omega_L = \omega_A$, we must have

$$\omega_L^0 + kv(x) = \omega_A^0 + \mu B(x)/\hbar.$$

By using Eq. (9.4), we get the expression of $B(x)$:

$$B(x) = \frac{\hbar}{\mu}\left(\omega_L^0 - \omega_A^0 + k\sqrt{v_i^2 - 2a_0 x}\right).$$

The magnetic field profile thus has a parabolic shape.

(d) For the Zeeman effect to compensate the Doppler effect associated with a velocity change Δv, the magnetic field must vary by

$$\Delta B = \frac{\hbar k\,\Delta v}{\mu_B} \simeq 46 \times 10^{-3}\,\mathrm{T}.$$

This can be realized in a relatively easy way in the laboratory, e.g. using a tapered solenoid.

(3) **Cooling effect.**

(a) The acceleration $a < a_0$ and the initial and final velocities (v_i and v_f, respectively) being fixed, the length ℓ of the slowing region is imposed by

$$v_f^2 = v_i^2 - 2a\ell. \tag{9.5}$$

Let us stress that, in order to have a constant acceleration $a < a_0$, the detuning $\omega_L - \omega_A$ must have a constant value, that we shall denote by δ_a.

(b) We first express $v(x)$ as a function of v_i, v_f, x and ℓ: we have $v(x) = \sqrt{v_i^2 - 2ax}$; but, from Eq. (9.5), $(v_f^2 - v_i^2)/\ell = -2a$. We thus get:

$$v(x) = \sqrt{v_i^2 + (v_f^2 - v_i^2)\frac{x}{\ell}}.$$

The magnetic field profile is imposed by the relation $\omega_L - \omega_A = \omega_L^0 + kv(x) - \omega_A^0 - 2\pi\mu B(x) = \delta_a$. We thus have:

$$B(x) = \frac{\hbar}{\mu}\left(\omega_L^0 - \omega_A^0 - \delta_a + k\sqrt{v_i^2 + (v_f^2 - v_i^2)\frac{x}{\ell}}\right)$$

$$= B_0 + \frac{\hbar k}{\mu}\sqrt{v_i^2 + (v_f^2 - v_i^2)\frac{x}{\ell}}.$$

(c) Let us study, as a function of its initial velocity, the evolution of an atom in the slower.

- We first consider a particle whose initial velocity is significantly[7] higher than v_i. Initially, it is thus not resonant with the slowing laser, and, consequently, is not decelerated. While it travels along the slower (at a constant velocity $v > v_i$), the detuning from resonance still increases (since, at a given position $x > 0$, the resonance condition is fulfilled for atoms having a velocity *smaller* than v_i), and thus this atom will never reach the resonance. Such a particle is thus not affected by the slower, and will exit it with its initial velocity. In phase space (x, v), the trajectory of such a particle is simply a straight, horizontal line (curve (i) in Fig. 9.6).

- Consider now a particle with an initial velocity between v_f and v_i. Initially, it is not resonant and thus it is not decelerated. But, during its propagation along the slower, it comes closer to resonance, and begins to "feel" the radiation pressure force. At a given position x, the radiation pressure force and the atomic velocity are identical to the ones for an atom that had entered the slower with a velocity v_i. The atom thus goes on being decelerated until it reaches a velocity v_f (see curve (ii) in Fig. 9.6).

---

[7]i.e. which differs from v_i by more than a few $\bar{\Gamma}/k$, a quantity which is numerically much smaller than the typical velocities that are considered here, see question 2(b).

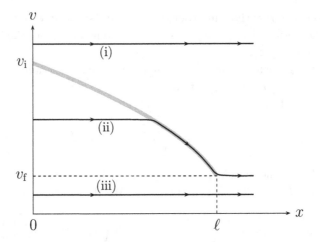

Fig. 9.6   *The slowing process seen in phase space (x, v). The parabolic shaded area corresponds to points in phase space where the radiation pressure force has non-negligible values. An atom with an initial velocity higher than v_i is never resonant, and thus is not decelerated: see trajectory (i). On the contrary, an atom whose initial velocity lies in between v_f and v_i will become resonant with the laser at some point $0 < x < \ell$, and will slow down in such a way that it stays in resonance; its final velocity will thus be close to v_f: see curve (ii). Finally, an atom with an initial velocity smaller than v_f is never resonant, and thus not decelerated: see trajectory (iii).*

- Finally, a particle with an initial velocity significantly smaller than v_f is never resonant, it thus keeps its velocity (see curve (iii) in Fig. 9.6).

(d) From the discussion above, the distribution $p(v)$ of the atomic velocities in the beam at $x = \ell$ has the behavior shown in Fig. 9.7: for $v > v_i$ or $v < v_f$, we have $p(v) = p_0(v)$, where $p_0(v)$ is the velocity distribution in the beam in the absence of slower (dashed line). Since all the particles with an initial velocity $v \in [v_f, v_i]$ have a final velocity on the order of v_f, a "hole" is dug in the velocity distribution, and the corresponding atoms are found in a large peak centered at v_f (solid line). In practice, the width of this peak is determined by the way the atoms fall out of resonance in $x = \ell$ (indeed, one cannot have a very fast variation of $B(x)$ at the output of the slower), and is of a few meters per second. Therefore, in a Zeeman slower, one does not only *slow down* the beam (i.e. decrease the mean velocity), as the name "slower" implies, but one also observes *cooling*: the velocity dispersion strongly decreases.

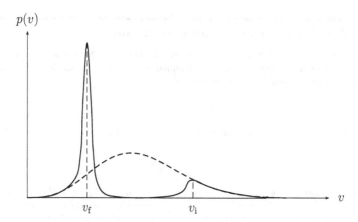

Fig. 9.7   *Schematic representation of the atomic velocity distribution in the beam at the input (dashed line) and at the output (solid line) of the slower. The atoms whose initial velocity was between v_f and v_i have a final velocity almost equal to v_f.*

♦ **Remark.** It is possible, while having a fixed magnetic field profile $B(x)$, to adjust the final velocity v_f of the atoms by simply modifying the laser frequency ω_L^0. Indeed, if one changes ω_L^0 in such a way that the atoms that are resonant at $x = 0$ are slower than v_i, these atoms will be able to follow the resonance condition without any problem (the acceleration required in order to decelerate them over a distance ℓ is smaller than a). The final velocity of these atoms is then smaller than v_f. In the opposite case (the atoms that are resonant at $x = 0$ have a velocity higher than v_i), one must have an acceleration higher than a, which is possible only as long as the required acceleration remains smaller than a_0 (otherwise, the magnetic field profile is too "compact" for decelerating such fast atoms).

The device described in this problem was realized for the first time[8] by the group of the American physicist W.D. Phillips at the beginning of the eighties. This was a breakthrough in atomic physics, allowing for new experiments to be performed thanks to the unprecedented low velocities achieved in this way. Zeeman slowers are nowadays widely used in atomic physics experiments as sources of slow atoms. This research was recognized by the Nobel Committee, which awarded the Nobel Prize in Physics to W.D. Phillips in 1997 (jointly with C. Cohen-Tannoudji and S. Chu, for their achievements concerning laser cooling of neutral atoms).

---

[8]W.D. Phillips and H. Metcalf, *Laser Deceleration of an Atomic Beam*, Phys. Rev. Lett. **48**, 596 (1982).

Reminder: Photon energy and momentum

An electromagnetic plane wave of wave vector $\boldsymbol{k}$ (and thus having a wavelength $\lambda = 2\pi/k$) and frequency ν (i.e. of angular frequency $\omega = 2\pi\nu$) is associated, quantum-mechanically, with *photons* having an energy

$$E = \hbar\omega = h\nu, \tag{9.6}$$

where $h = 2\pi\hbar$ is Planck's constant, and a linear momentum

$$\boldsymbol{p} = \hbar\boldsymbol{k} = \frac{h}{\lambda}\boldsymbol{u}, \tag{9.7}$$

where $\boldsymbol{u}$ is a unit vector along $\boldsymbol{k}$. Equations (9.6) and (9.7) are called, respectively, Einstein's relation and de Broglie's relation. From these equations, it appears that the dispersion relation for electromagnetic waves $\omega = ck$ is equivalent to the energy-momentum relation $E = pc$ valid for ultrarelativistic particles.

9.3 Doppler cooling **

Using lasers, it is possible to prepare clouds of atoms containing typically 10^9 particles at a temperature well below one millikelvin! This type of cooling is obtained by illuminating the atoms by laser beams that are close to resonance with a transition between two internal levels of the atom. In these conditions, the atom which absorbs a photon also acquires its linear momentum[9], which results in a force, called the *radiation pressure force*.

In this problem, we shall study a simple model of this force, in order to understand how it can damp the atomic motion. Including the heating effects linked to the unavoidable recoil which the atom experiences when a photon is reemitted, we will finally deduce the minimal temperature that one can reach in this way — the so-called *Doppler temperature*.

Radiation pressure

An atom in a quasi-resonant light field can be modeled as a driven harmonic oscillator. We consider an alkali atom, with a single valence electron. This electron, with charge q and mass m, is assumed to be harmonically bound to the nucleus, with an angular frequency ω_0 and a damping coefficient Γ. The

---

[9]See reminder on page 188.

driving term induced by the laser[10] arises from the oscillating electric field $E = E_0 \cos \omega t$. The displacement $z(t)$ of the electron around its equilibrium position is thus the solution of the equation:

$$\ddot{z} + \Gamma \dot{z} + \omega_0^2 z = \frac{q}{m} E_0 \cos \omega t. \tag{9.8}$$

After the transient regime, the steady state solution of Eq. (9.8) has the form $z(t) = z_0 \cos(\omega t - \phi)$.

(1) Determine the amplitude z_0 and phase ϕ as a function of q, m, E_0, ω_0, δ and Γ, assuming that the detuning $\delta \equiv \omega - \omega_0$ and the damping Γ are small compared to the resonance frequency ($|\delta|, \Gamma \ll \omega_0$).

(2) Give the expression of the average power $\langle P \rangle$ absorbed by the atom as a function of q, m, E_0, δ and Γ.

(3) To find the expression of the force F exerted on the atom by an incident quasi-monochromatic electromagnetic plane wave, we use a corpuscular interpretation of light. In this picture, the force F is obtained by multiplying the photon absorption rate r by the momentum $\hbar k$ of the incident photons. This force, referred to as the *radiation pressure force*, pushes the atom in the propagation direction of the laser. Calculate r, and then F, as a function of the illumination intensity $I = \varepsilon_0 c E_0^2 / 2$ and of the *saturation intensity* $I_s = \varepsilon_0 m c \Gamma^2 \hbar \omega / q^2$.

(4) Calculate numerically the acceleration a of a rubidium atom illuminated by a resonant beam at wavelength $\lambda = 780$ nm, for an intensity $I = I_s/2 \simeq 0.8$ mW $\cdot$ cm^{-2}.

Doppler damping

In order to cool a gas one must decrease the energy of the atoms. It is therefore necessary to have a velocity-damping mechanism, i.e. a force which depends on the atomic velocity. This dependence arises here from the Doppler effect at work in the interaction between an atom and the laser beam which illuminates it. In one dimension (along the x-axis), and for a laser field propagating along the negative direction with angular frequency ω, an atom of velocity $v_x > 0$ "sees" the laser oscillate at the frequency $\omega' = \omega(1 + v_x/c)$ due to the Doppler effect.

(1) Comment on the expression for ω'. In the following, we thus calculate the force exerted on the atom, taking the Doppler shift into account,

---

[10]The laser is modeled as a plane wave propagating along the x direction and linearly polarized along the z axis.

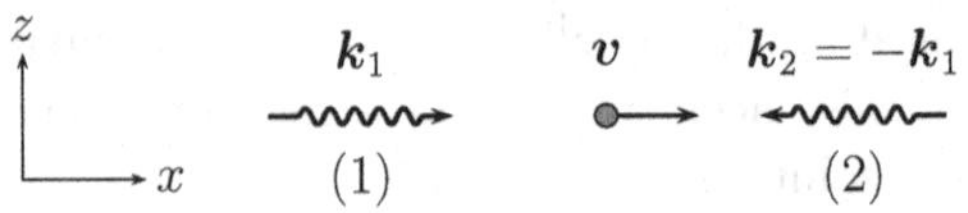

Fig. 9.8 *An atom with velocity $v_x = v > 0$ in the field of two counter-propagating laser waves.*

by replacing the detuning $\delta = \omega - \omega_0$ by $\delta + kv_x$.

(2) Irradiating an atom with a laser beam propagating in the direction opposite to its motion allows for a decrease in velocity. However, if the radiation pressure force continues to act on the atom, it will end up by accelerating it along the direction of propagation of the beam. On the contrary, we look for a situation where an atom at rest does not experience any force. We thus consider the configuration shown in Fig. 9.8, with two identical laser beams (same intensity I and frequency ω).

We assume that the two beams act independently on the atom. Calculate the resulting force exerted on an atom with velocity v_x. Show that for small velocities, i.e. for $k|v_x| \ll |\delta|, \Gamma$, this force can be written as $F = (r_+ + r_-)\hbar k = -\alpha v_x$. Give the explicit expression of the coefficient α.

(3) Show that the energy E of an atom with mass M, submitted to the damping force $F = -\alpha v_x$, decreases exponentially with time. Calculate the characteristic time constant τ of this decay. For the case $I = I_s/10$ and $\delta = -\Gamma/2$, evaluate τ numerically for a rubidium atom.

(4) Discuss qualitatively the energy and entropy balance in Doppler cooling, assuming that the atom reemits a photon at the same frequency as the one it absorbs in the frame where it is at rest.

Heating due to spontaneous emission

The longer the damping force acts, the more the atomic velocity can be reduced. However, one has to face a competing heating mechanism which prevents one from reaching arbitrarily low temperatures. Indeed, quantum mechanically, the atom interacts with the laser fields by absorption and emission of photons. When the atom absorbs a photon, it jumps from its internal ground state to an excited state. The latter has a finite lifetime Γ^{-1}. The atom jumps back to the ground state by spontaneous emission of a photon; this emission takes place in a random direction. As a result of these elementary processes, the atom undergoes a random walk in velocity space.

For the one-dimensional situation that we consider here, a consequence of this random walk is that the r.m.s. momentum increases according to

$$\frac{d\langle p^2 \rangle}{dt} = (r_+ + r_-)(\hbar k)^2.$$

(1) Deduce the corresponding heating in the limit of low velocities:

$$\left(\frac{dE}{dt}\right)_{\text{heating}} = \frac{1}{2M}\frac{d\langle p^2 \rangle}{dt}.$$

(2) In steady state, the cooling and heating processes compensate for each other:

$$\left(\frac{dE}{dt}\right) = \left(\frac{dE}{dt}\right)_{\text{cooling}} + \left(\frac{dE}{dt}\right)_{\text{heating}} = 0.$$

Find the equilibrium temperature by using the equipartition theorem in one dimension.

(3) Show that this temperature reaches a minimum when $\delta = -\Gamma/2$ and give the explicit expression of this temperature as a function of $\hbar$, Γ and k_B. Calculate its numerical value for rubidium atoms, as well as the corresponding r.m.s. velocity.

Data for the numerical calculations.

For a rubidium atom, one has:

$$M = 1.45 \times 10^{-25} \text{ kg},$$
$$\Gamma^{-1} = 2.7 \times 10^{-8} \text{ s},$$
$$\lambda = 780 \times 10^{-9} \text{ m}.$$

Solution

Radiation pressure

(1) By substituting the steady-state solution $z = z_0 \cos(\omega t - \phi)$ into Eq. (9.8), one finds, in the limit $|\delta|, \Gamma \ll \omega_0$:

$$z_0 = \frac{q}{m}\frac{E_0}{2\omega_0}\frac{1}{(\delta^2 + \Gamma^2/4)^{1/2}} \quad \text{and} \quad \phi = -\tan^{-1}\left(\frac{\Gamma}{2\delta}\right).$$

(2) The instantaneous power P is the product of the force by the velocity:

$$P = F\dot{z} = -qE_0\omega z_0 \cos(\omega t)\sin(\omega t - \phi).$$

The average, calculated over an oscillation period, is thus

$$\langle P \rangle = \frac{1}{2}qE_0\omega z_0 \sin\phi.$$

The relationship $\sin\phi = \tan\phi/(1 + \tan^2\phi)^{1/2}$ allows us to rewrite this average as a Lorentzian in the detuning δ:

$$\langle P \rangle = \frac{\Gamma\omega^2 q^2 E_0^2}{2m\omega_0^2}\frac{1}{4\delta^2 + \Gamma^2} \simeq \frac{q^2}{2m}\frac{\Gamma}{4\delta^2 + \Gamma^2}E_0^2.$$

where we have used $\omega \simeq \omega_0$.

(3) Each photon has an energy $\hbar\omega$, therefore the photon absorption rate is $r = \langle P \rangle/\hbar\omega$. The force F_x exerted on the atom is thus:

$$F_x = r\hbar k = \frac{I}{I_s}\frac{1}{1 + 4\delta^2/\Gamma^2}\Gamma\hbar k.$$

(4) For $\delta = 0$, we have $a = F/m = I\Gamma\hbar k/(mI_s) = 1.1 \times 10^5$ m·s^{-2}. This is a very large acceleration, on the order of 10^4 times as large as the one due to gravity. We note that from the expression above, it seems that the force increases without bound when the intensity increases. In fact, our simple model is valid only at low intensity, i.e. for $I \ll I_s \simeq 1.6$ mW·cm^{-2}, in a regime where the contribution of stimulated emission is negligible. Indeed, at high intensity, one observes saturation effects, which limit the radiation pressure force to the maximal value $\Gamma\hbar k/2$. Only a quantum-mechanical model of the atom can describe saturation correctly.

Doppler damping

(1) To understand the physical meaning of the expression $\omega' = \omega + kv_x$ of the Doppler shift, let us consider a simple analogy. We imagine that an apparatus (a source) throws $\omega/(2\pi)$ particles per second to a receiver. The particles are all thrown with the same velocity c. If the receiver does not move with respect to the source, it will receive $\omega/(2\pi)$ particles every second. However, if it moves towards the source with velocity $v_x > 0$, it will receive particles more frequently; using the non-relativistic addition of velocities, one finds that the number of particles received per second is $\omega'/(2\pi)$. Similarly, if it moves away from the source, the apparent frequency is decreased.

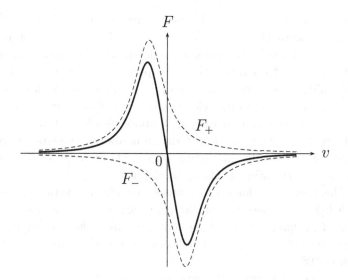

Fig. 9.9   *The cooling force $F(v)$ (solid line) is the sum of the contributions F_+ and F_- of the two counterpropagating waves (dashed line).*

Note that the expression $\omega' = \omega + kv_x$ is only accurate to first order in v_x/c, where c is the speed of light. This is the case for atomic velocities, since they are at most on the order of the thermal velocities at room temperature, i.e. $v_x \sim 300\ \mathrm{m \cdot s^{-1}} \ll c$.

(2) The forces read (Fig. 9.8):

$$F_\pm = r_\pm \hbar k = \pm \frac{I}{I_s} \frac{1}{1 + 4(\delta \mp kv_x)^2/\Gamma^2} \Gamma \hbar k$$

Assuming that the forces from the two beams act independently, the resulting force is:

$$F = F_+ + F_- = \frac{I}{I_s}\Gamma\hbar k \left(\frac{1}{1 + 4(\delta - kv_x)^2/\Gamma^2} - \frac{1}{1 + 4(\delta + kv_x)^2/\Gamma^2} \right).$$

The curve $F(v)$ has a dispersive shape (see Fig. 9.9). For low velocities, one can expand the expression of the force to first order in v_x. We find $F \simeq -\alpha v_x$ with

$$\alpha = -16 \frac{I}{I_s} \hbar k^2 \frac{\delta/\Gamma}{(1 + 4\delta^2/\Gamma^2)^2}. \tag{9.9}$$

We notice that there is a damping only in the case of a negative detuning $\delta = \omega - \omega_0 < 0$. The interpretation of this result is clear: when an atom moves with velocity $v_x = v > 0$ towards the right, it "sees", by the

Doppler effect, the beam (2) closer to resonance ($\omega' = \omega + kv \simeq \omega_0$) than the beam (1); it therefore preferentially absorbs photons from beam (2). Its velocity decreases as it absorbs the linear momentum of the absorbed photons. Since the photon reemission has the same probability of occurring into two opposite directions, it does not contribute, on average, to the velocity change[11]. If the atom has a velocity $v_x = v < 0$, the same reasoning as above also leads to a decrease in velocity. In summary, for $\delta < 0$, the atom always undergoes a decrease in velocity due to the interaction with the laser fields. In the opposite case ($\delta > 0$) the atom would be accelerated.

The viscous-like force which arises from the interaction of the atom with light is responsible for the name *optical molasses* given to this type of medium made of laser-cooled atoms. The first experimental observation[12] of such optical molasses was done by the group of Steven Chu in 1985.

(3) We just need to rewrite the expression of dE/dt

$$\frac{dE}{dt} = \boldsymbol{v} \cdot M \frac{d\boldsymbol{v}}{dt} = -\alpha v^2 = -\frac{2\alpha}{M} E. \qquad (9.10)$$

With the numerical values given above, we calculate $\tau = M/(2\alpha) \simeq 50 \ \mu s$ for rubidium atoms. The damping is thus very fast.

(4) It is worth having a closer look at the balance of energy and entropy in laser Doppler cooling.

In order to understand energy balance, consider the case of an atom moving in the x direction with a velocity $\boldsymbol{v} = v\boldsymbol{e}_x$.

(a) If the atom absorbs a photon from wave (2) (see Fig. 9.8), it reemits a photon $\omega + 2kv$ in the $+x$ direction, and ω in the $-x$ direction since the atom reemits a photon at the same frequency as the one it absorbs in a frame where it is at rest (this is true for low intensities). On average, the energy of the reemitted photon is larger than the energy of the absorbed one, even after angular integration over the direction of the reemitted photon.

(b) If the atom absorbs a photon from wave (1), it reemits a photon $\omega - 2kv$ in the $-x$ direction and ω in the $+x$ direction. On the average the energy of the reemitted photon is smaller than the energy of the absorbed one.

---

[11]Note that the reemission induces a random walk in velocity space. We shall see in the next part that this sets a limit temperature for the gas.

[12]S. Chu, L. Hollberg, J. Bjorkholm, A. Cable, and A. Ashkin, *Three-dimensional viscous confinement and cooling of atoms by resonance radiation pressure*, Phys. Rev. Lett. **55**, 48 (1985).

If processes (a) and (b) have the same rate, then on average there is no change of energy of the radiation field, and thus of the atom. Since the laser is red detuned, however, process (a) is more frequent than process (b) when the atom velocity is positive ($v > 0$), so that the energy of the reemitted photon is on average larger than the energy of the absorbed one. We conclude that the energy of the radiation field increases while the energy of the atoms decreases.

Cooling a cloud of atoms results in a decrease of the entropy of the atoms. However, one must not forget that to achieve Doppler laser cooling, photons are absorbed from the laser beam. In this process, photons from the laser beam, which has a low entropy, are transformed into fluorescence photons emitted in all possible directions. The fluorescence field is a disordered system with a high entropy. The entropy of the radiation field thus increases while the entropy of the atoms decreases. In this context, the second law of thermodynamics states that the entropy for the total system {atom + field} increases.

Heating due to spontaneous emission

(1) When v is small,

$$\frac{\mathrm{d}\langle p^2 \rangle}{\mathrm{d}t} = 2(r_+ + r_-)(\hbar k)^2 = \frac{4\Gamma I/I_\mathrm{s}}{1 + (2\delta/\Gamma)^2}(\hbar k)^2.$$

We deduce the contribution to the "heating" of the atomic ensemble:

$$\left(\frac{\mathrm{d}E}{\mathrm{d}t}\right)_\mathrm{heating} = \frac{1}{2M}\frac{\mathrm{d}\langle p^2 \rangle}{\mathrm{d}t} = \frac{(\hbar k)^2}{2M}\frac{4\Gamma I/I_\mathrm{s}}{1 + (2\delta/\Gamma)^2}.$$

(2) In steady state we have:

$$\left(\frac{\mathrm{d}E}{\mathrm{d}t}\right)_\mathrm{heating} = -\left(\frac{\mathrm{d}E}{\mathrm{d}t}\right)_\mathrm{cooling} = \frac{E}{\tau},$$

where we have used Eq. (9.10), and where $\tau = M/(2\alpha)$. The equipartition theorem states that the average energy $\langle E \rangle$ per particle that is associated to one degree of freedom[13], for non-interacting particles, in thermal equilibrium at temperature T, and whose energy depends quadratically on the dynamical variable[14], is equal to $\langle E \rangle = k_\mathrm{B}T/2$. We thus have, using the expression (9.9) of α:

$$\langle E \rangle = \frac{1 + (2\delta/\Gamma)^2}{-2\delta/\Gamma}\frac{\hbar\Gamma}{8} = \frac{k_\mathrm{B}T}{2}.$$

---

[13] Here, this degree of freedom is that of the velocity v_x along the x-axis.
[14] This is indeed the case here as the kinetic energy reads $mv_x^2/2$.

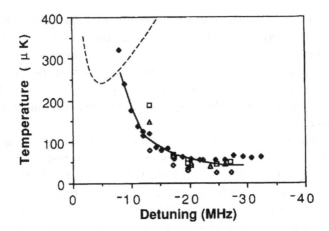

Fig. 9.10 *Comparison between the observed temperature (points) of atoms in a three-dimensional optical molasses and the Doppler prediction detailed in this exercise (dashed line). These experiments were performed with sodium atoms, for which $\Gamma = 2\pi \times 10$ MHz, yielding a Doppler temperature of 240 μK. Figure adapted (with permission) from Phys. Rev. Lett. **61**, 169 (1988). Copyright American Physical Society.*

(3) The minimum temperature $T_{\min}$ is obtained by minimizing the function $\langle E \rangle(\delta)$ with respect to δ. We find $T_{\min} = \hbar\Gamma/(2k_{\rm B}) \simeq 140\ \mu$K, for a detuning $\delta = -\Gamma/2$. In terms of rms velocity, such a temperature corresponds to $v \sim (k_{\rm B}T/M)^{1/2} \simeq 12\ {\rm cm \cdot s}^{-1}$. Compared to a vapor at room temperature, the thermal velocity is divided by one thousand, and the temperature by one million! Laser cooling is thus a very efficient technique.

Laser cooling of neutral atoms was first proposed theoretically in 1975[15] by Arthur Schawlow and Ted Hänsch, who were awarded the Nobel Prize in Physics in 1981 and 2005, respectively. Let us note that the estimate of the limit temperature in this publication turned out to be wrong, actually greatly overestimated. This might relieve the reader who would have found the problem hard to solve! In 1976, a group of Soviet physicists established the correct expression for the limit temperature[16]. However, experiments performed by the group of W.D. Phillips in 1988[17] revealed temperatures that were much lower

[15]T. Hänsch and A. Schawlow, *Cooling of gases by laser radiation*, Opt. Commun. **13**, 68 (1975).

[16]V.S. Letokhov, V.G. Minogin and B.D. Pavlik, *Cooling and trapping of atoms and molecules by a resonant laser field*, Opt. Commun. **19**, 72 (1976).

[17]P.D. Lett, R.N. Watts, C.I. Westbrook, W.D. Phillips, P.L. Gould, and H.J. Metcalf,

than the ones deduced from the Doppler cooling model (see Fig. 9.10)!

Claude Cohen-Tannoudji, Steven Chu, and William Phillips were awarded the Nobel Prize in Physics in 1997 for their pioneering work in laser cooling of neutral atoms. Among other major contributions, the first two explained theoretically the existence of temperatures lower than the Doppler limit (the physical reason being that (i) the atoms used in experiments have a much richer internal structure than the simple two-level approximation used in the Doppler cooling model and (ii) that the local polarization of the electromagnetic field resulting from the superposition of the two counter-propagating travelling waves is not uniform).

9.4 Time-of-flight thermometry **

In this exercise, we study the so-called *time-of-flight* method that is used to measure the temperature of trapped gases of ultracold atoms.

We consider a gas of N identical atoms of mass m that are confined in a harmonic trap of angular frequency ω_0. We restrict the discussion to a one-dimensional situation, although the generalization to three dimensions is straightforward. The initial distribution in position and in velocity is given by the Boltzmann law (see reminder on page 212):

$$P_0(x, v_x) = \frac{Nm\omega_0}{2\pi k_B T} \, e^{-m\omega_0^2 x^2/(2k_B T)} \, e^{-mv_x^2/(2k_B T)}. \tag{9.11}$$

The confinement is abruptly removed at $t = 0$, and the mean square radius $\langle x^2 \rangle(t)$ of the cloud is measured as a function of time, while it is falling due to gravity. Calculate $\langle x^2 \rangle(0)$, and show how it is possible to extract the temperature T from the knowledge of $\langle x^2 \rangle(t)$[18]. Table 9.1 gathers experimental data from such a time-of-flight measurement performed on a laser cooled cloud of rubidium 87 atoms ($m = 1.45 \times 10^{-25}$ kg). Deduce the temperature T of the cloud (we recall the value of the Boltzmann constant $k_B = 1.38 \times 10^{-23}$ J $\cdot$ K^{-1}).

Observation of atoms laser-cooled below the Doppler limit, Phys. Rev. Lett. **61**, 169 (1988).

[18] As the reader will show easily, gravity does not influence this result; thus it is not taken into account in the following.

Table 9.1   *Results of a time-of-flight experiment.*

$t/$ms	2	3	4	5	6	7	8	9
$\sqrt{\langle x^2 \rangle}/\mu$m	268	305	344	387	453	502	583	638

Solution

By definition, initially, the mean square radius of the cloud is given by the average value of x^2, weighted by the initial phase-space distribution $P_0(x, v_x)$:

$$\langle x^2 \rangle(0) = \frac{\displaystyle\iint x^2 P_0(x, v_x)\, \mathrm{d}x\, \mathrm{d}v_x}{\displaystyle\iint P_0(x, v_x)\, \mathrm{d}x\, \mathrm{d}v_x} = \frac{k_B T}{m\omega_0^2}.$$

As intuitively expected, the lower the temperature and the stronger the strength of the confining potential, the smaller the size of the cloud. In principle, measuring this initial size and the trapping frequency of the harmonic confinement is sufficient to determine the temperature. However, measuring the cloud size *in situ* is often impractical due to the finite resolution of the imaging system, and measuring accurately the trapping frequency is also challenging. The time-of-flight method avoids this inconvenience, as we shall see below. Note that the size of the cloud is independent of the number of atoms; this is due to the fact that we have assumed that the particles do not interact[19].

To calculate the average quadratic size $\langle x^2 \rangle(t)$, we propose two different methods.

- The first one simply consists in using the evolution of each particle separately. Since no force is exerted on the atoms after the switching off of the trapping potential, each atom i evolves freely and thus: $x_i(t) =$

---

[19] One can compare this result to the one obtained in Problem 8.1 about trapped ions, for which the size of the cloud scales as $N^{1/3}$ with the number N of ions, because of Coulomb repulsion.

$x_i(0) + v_i(0)t$. Consequently, we get for $\langle x^2 \rangle(t)$:

$$\langle x^2 \rangle(t) = \frac{1}{N} \sum_{i=1}^{N} x_i^2(t)$$

$$= \frac{1}{N} \sum_{i=1}^{N} [x_i(0) + v_i(0)t]^2$$

$$= \frac{1}{N} \sum_{i=1}^{N} [x_i^2(0) + 2x_i(0)v_i(0)t + v_i^2(0)t^2]$$

$$= \langle x^2 \rangle(0) + 2\langle xv_x \rangle(0)t + \langle v_x^2 \rangle(0)t^2.$$

One can easily check that for the distribution (9.11), we have $\langle xv_x \rangle(0) = 0$ and $\langle v_x^2 \rangle(0) = k_B T/m$. The first relation means that there are no initial correlations between positions and velocities of the particles. This can be seen directly on the expression of the initial distribution $P_0(x, v_x)$, since it is the product of two independent distributions, one for the positions and the other one for the velocities:

$$P_0(x, v_x) = N \sqrt{\frac{m}{2\pi k_B T}} e^{-mv_x^2/(2k_B T)} \times \sqrt{\frac{m\omega_0^2}{2\pi k_B T}} e^{-m\omega_0^2 x^2/(2k_B T)}.$$

We finally obtain:

$$\langle x^2 \rangle(t) = \frac{k_B T}{m} \left(\frac{1}{\omega_0^2} + t^2 \right).$$

We conclude that by plotting $\langle x^2 \rangle$ as a function of t^2 one obtains a straight line, whose slope $k_B T/m$ yields the temperature of the cloud.

• The other, equivalent method that we propose relies on the use of the Dirac delta function. The first step consists in establishing the expression for the atomic density $n(x,t)$ as a function of time. By definition,

$$n(x,t) = \int_{-\infty}^{\infty} \int_{-\infty}^{\infty} P_0(x_0, v_x)\delta(x - x_0 - v_x t)\mathrm{d}x_0 \mathrm{d}v_x$$

$$= \frac{1}{t} \int_{-\infty}^{\infty} P_0 \left(x_0, \frac{x - x_0}{t} \right) \mathrm{d}x_0$$

$$= N \sqrt{\frac{m\omega_0^2}{2\pi k_B T} \frac{1}{1 + \omega_0^2 t^2}} \exp \left(-m\omega_0^2 \frac{x^2}{2k_B T(1 + \omega_0^2 t^2)} \right).$$

We infer the expression for $\langle x^2 \rangle(t)$:

$$\langle x^2 \rangle(t) = \frac{1}{N} \int_{-\infty}^{\infty} x^2 n(x,t)\mathrm{d}x = \frac{k_B T}{m} \left(\frac{1}{\omega_0^2} + t^2 \right).$$

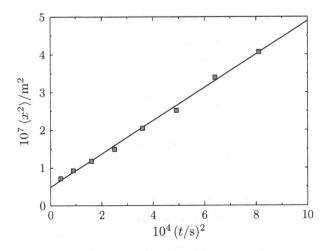

Fig. 9.11 *Mean square radius* $\langle x^2 \rangle$ *of the cloud of atoms in a time-of-flight experiment as a function of* t^2.

With the experimental data points of Table 9.1, we plot the mean square radius $\langle x^2 \rangle$ of the cloud as a function of t^2 (see Fig. 9.11). We obtain a straight line of slope 4.4×10^{-3} m $\cdot$ s^{-2}. We deduce the value of the velocity dispersion $\sqrt{\langle v_x^2 \rangle} = 6.6$ cm $\cdot$ s^{-1}, and the temperature $T = m\langle v_x^2 \rangle / k_B = 46$ μK! This very low temperature is obtained through laser cooling techniques. We point out that such ultracold gases have a small number of atoms, on the order of a few billion. The reader interested in this subject can read the Nobel lectures of Claude Cohen-Tannoudji, William D. Phillips, and Steven Chu (http://nobelprize.org/nobel_prizes/physics/laureates/1997/). Problem 9.3 on Doppler cooling gives an introduction to the physics of laser cooling.

9.5 Linear and parametric heating ***

In this exercise, we consider a gas of N identical atoms of mass m that are magnetically trapped. Indeed, the magnetic moment of the atom can be used to ensure its confinement with an appropriate configuration of static magnetic fields. In this situation, the atoms evolve according to the laws of classical physics in an external potential U provided by the interaction between the magnetic moment and the magnetic field, and undergo elastic collisions between themselves.

In practice, magnetic fields are generated by electric currents running through coils. The experimental observation is the following: depending on the quality of the power supplies that provide the currents, the temperature of the gas (measured by a time-of-flight technique, see Problem 9.4) either remains constant, or grows with time, exponentially or linearly. Our aim is to understand the origin of these various behaviors.

To simplify the analysis, we only consider the one-dimensional situation and, in addition, we assume that the confinement is harmonic, with an angular frequency ω_0 .

(1) The increase of temperature corresponds to an increase of the mean energy of the gas. Can a time-independent potential explain the experimentally observed increase of temperature? Comment.

(2) The variations with time of the trapping potential that are considered in the following are assumed to be small so that their effect can be described either (i) by a modification of the strength of the trapping potential (i.e. by a modification of the angular frequency), or (ii) by a random motion of the position of the bottom of the confining potential. In order to account for both effects, the potential considered in the following reads:

$$U_{\text{trap}}(x,t) = \frac{1}{2}m\omega_0^2[1 + \epsilon(t)][x - a(t)]^2,$$

where $|\epsilon(t)| \ll 1$ accounts for the time variation of the angular frequency, and $a(t)$ for the time-dependent position of the bottom of the trap. Show that for both cases the equation of motion of the atoms in the time-dependent potential can be put in the form:

$$\ddot{x} + \omega_0^2 x = f(t). \tag{9.12}$$

Hint: use an expansion to first order.

Give, for each kind of time-dependent term in the potential, the explicit expression of the source term $f(t)$.

(3) Show that:

$$x(t) = x(0)\cos(\omega_0 t) + \frac{\dot{x}(0)}{\omega_0}\sin(\omega_0 t) + \frac{1}{\omega_0}\int_0^t \sin\left[\omega_0(t - \tau)\right]f(\tau)\,\mathrm{d}\tau$$

is a solution of Eq. (9.12).

(4) Give the explicit expression of $\langle \dot{E} \rangle$, the time derivative of the total energy per particle, averaged over one oscillation period, as a function of $\dot{x}(t)$ and $f(t)$.

(5) In order to capture the role played by the source term $f(t)$, we carry out in the following a first-order expansion of the exact solution with $f(t)$ as a small parameter. The solution can therefore be written as

$$x(t) = x_0(t) + x_1(t) + \dots$$

where $x_0(t)$ is solution of Eq. (9.12) in the absence of the source term $f(t)$. Give the explicit expression for $x_0(t)$, $\dot{x}_0(t)$, $x_1(t)$, and $\dot{x}_1(t)$.

(6) We consider in this question that the strength of the potential remains constant with time: $\epsilon(t) = 0$. Show that $\langle \dot{E} \rangle = \Gamma$, and give the expression of the constant Γ. Comment.

(7) We now consider the opposite situation, for which the bottom of the trap always remains at the same position: $a(t) = 0$. Establish the integral expression of $G(t) \equiv \dot{x}_0(t)x_1(t) + \dot{x}_1(t)x_0(t)$. Deduce from it that $\langle \dot{E} \rangle = \Gamma' \langle E \rangle$. Give the expression of the constant Γ'. Conclude.

Reminder: The value of the Gauss integral is[20]

$$\int_{-\infty}^{\infty} e^{-u^2} \, du = \sqrt{\pi}.$$

Solution

An atom with a permanent magnetic moment is sensitive to the magnetic field. The potential energy associated with the interaction between the magnetic moment $\boldsymbol{\mu}$ and the field $\boldsymbol{B}(\boldsymbol{r}, t)$ is of the form $U = -\boldsymbol{\mu} \cdot \boldsymbol{B}(\boldsymbol{r}, t)$. If the magnetic moment is anti-parallel to the local direction of the magnetic field (this is the case for properly polarized atoms), the potential energy reads $U(\boldsymbol{r}, t) = \mu \|\boldsymbol{B}(\boldsymbol{r}, t)\| = \mu B(\boldsymbol{r}, t)$. The atoms are thus trapped in the vicinity of a local minimum of the modulus of the magnetic field. Note that, *a priori*, the reverse choice, where the atoms would have their magnetic moments parallel to the magnetic field, would in principle allow for trapping atoms in the vicinity of a maximum of the modulus of the magnetic field. However, according to Maxwell's equations, such a maximum in a three dimensional space cannot exist in a current-free region (i.e. outside the sources that generate the magnetic field). This result is known as Wing's theorem[21].

---

[20] See the reminder on page 179 for the demonstration.

[21] For a demonstration, see e.g. L. Pitaevskii and S. Stringari, *Bose-Einstein condensation*, Oxford University Press (2003), section 9.3.

(1) The temperature of the gas is proportional to the mean total energy of the gas, which remains constant in the presence of conservative forces. We thus have to assume that the trap potential varies as a function of time[22]. The source term related to the time-dependent force is responsible for the increase of temperature. It is worth stressing here the origin of the thermalization process. Indeed, for a gas confined in a box, thermalization occurs through both collisions between the particles themselves and collisions with the walls. In the trapped gas considered here, thermalization occurs only through collisions between particles. One cannot, strictly speaking, associate a temperature to the magnetic walls. However, if the magnetic walls move as a function of time, they can impart energy to the atoms. It is precisely this phenomenon that we study quantitatively in this exercise.

(2) The equation of motion reads:

$$m\ddot{x} = -\frac{\partial U_{\text{trap}}}{\partial x} = -m\omega_0^2[1 + \epsilon(t)][x - a(t)].$$

We infer the differential equation fulfilled by the position of each atom:

$$\ddot{x} + \omega_0^2 x = f(t) = \epsilon(t)\omega_0^2 x + \omega_0^2 a(t), \tag{9.13}$$

where we have neglected the second-order term scaling as $\epsilon(t)a(t)$. The first term of the right hand side of Eq. (9.13) accounts for the modulation of the strength of the confining potential, and the second term for the motion of the bottom of the trap.

(3) We can directly inject the proposed solution into Eq. (9.12), and check that it is a valid solution. Alternatively, one can obtain the result directly using, for instance, a matrix formalism. Let us introduce the vectors $\boldsymbol{X}$, $\boldsymbol{F}$ and the matrix M:

$$\boldsymbol{X} = \begin{pmatrix} x \\ \dot{x} \end{pmatrix}, \quad \boldsymbol{F}(t) = \begin{pmatrix} 0 \\ f(t) \end{pmatrix}, \quad \text{and} \quad \mathsf{M} = \begin{pmatrix} 0 & 1 \\ -\omega_0^2 & 0 \end{pmatrix}.$$

The second-order linear differential equation (9.13) can be recast in the form:

$$\frac{\mathrm{d}\boldsymbol{X}}{\mathrm{d}t} = \mathsf{M}\boldsymbol{X} + \boldsymbol{F}.$$

---

[22] Actually, one could envision other mechanisms yielding an increase in the temperature of the gas. Indeed, if inelastic collisions take place between the atoms, the energy stored in the internal degrees of freedom is transferred to the external degrees of freedom. This mechanism leads to an increase in the velocities of the atoms of the gas, and therefore in the total mean energy, i.e. in the temperature. Actually, the atoms cannot at the same time be in their ground state and have the correct polarization for being trapped magnetically, thus, they are prone to inelastic collisions. Note that the heating due to inelastic collisions can be easily distinguished from heating due to defects of the trapping potential, since the former depends on the atomic density.

The solution of this vectorial differential equation is simply given by:

$$\boldsymbol{X}(t) = \exp(\mathsf{M}t)\boldsymbol{X}(0) + \exp(\mathsf{M}t)\int_0^t \exp(-\mathsf{M}\tau)\boldsymbol{F}(\tau)\mathrm{d}\tau, \qquad (9.14)$$

where the vector $\boldsymbol{X}(0)$ contains the initial conditions, and where $\exp(\mathsf{M}t)$ stands for the exponential of the matrix $\mathsf{M}t$ (see the reminder on page 68). It remains to calculate the explicit expression of the matrix $\exp(\mathsf{M}t)$. The calculation of this quantity can be readily performed by diagonalizing the matrix M. However, for the specific case that we are studying here, a simplest strategy consists in noticing that $\mathsf{M}^2 = -\omega_0^2\mathsf{Id}$, where Id is the identity matrix (with diagonal elements equal to 1, and zeros everywhere else). As a result:

$$
\begin{aligned}
\exp(\mathsf{M}t) &= \sum_{n=0}^{\infty} \frac{t^n}{n!}\mathsf{M}^n \\
&= \sum_{n=0}^{\infty} \frac{t^{2n}}{(2n)!}\mathsf{M}^{2n} + \sum_{n=0}^{\infty} \frac{t^{2n+1}}{(2n+1)!}\mathsf{M}^{2n+1} \\
&= \sum_{n=0}^{\infty} \frac{(-1)^n(\omega_0 t)^{2n}}{(2n)!}\mathsf{Id} + \frac{1}{\omega_0}\sum_{n=0}^{\infty} \frac{(-1)^n(\omega_0 t)^{2n+1}}{(2n+1)!}\mathsf{M} \\
&= \cos(\omega_0 t)\mathsf{Id} + \frac{1}{\omega_0}\sin(\omega_0 t)\mathsf{M} \\
&= \begin{pmatrix} \cos(\omega_0 t) & \sin(\omega_0 t)/\omega_0 \\ -\omega_0\sin(\omega_0 t) & \cos(\omega_0 t) \end{pmatrix}.
\end{aligned}
$$

We finally obtain the explicit general expression for the solution of Eq. (9.14):

$$\boldsymbol{X}(t) = \begin{pmatrix} x(0)\cos(\omega_0 t) + \dfrac{\dot{x}(0)}{\omega_0}\sin(\omega_0 t) + \dfrac{1}{\omega_0}\displaystyle\int_0^t \sin\left[\omega_0(t-\tau)\right]f(\tau)\,\mathrm{d}\tau \\ -\omega_0 x(0)\sin(\omega_0 t) + \dot{x}(0)\cos(\omega_0 t) + \displaystyle\int_0^t \cos\left[\omega_0(t-\tau)\right]f(\tau)\,\mathrm{d}\tau \end{pmatrix}.$$

(4) Multiplying the differential equation fulfilled by $x(t)$ by the velocity $\dot{x}(t)$, one finds:

$$\dot{x}\ddot{x} + \omega_0^2\dot{x}x = \frac{\mathrm{d}}{\mathrm{d}t}\left(\frac{1}{2}\dot{x}^2 + \frac{1}{2}\omega_0^2 x^2\right) = \frac{1}{m}\frac{\mathrm{d}\mathcal{E}}{\mathrm{d}t} = \dot{x}f(t).$$

The average over the period $2\pi/\omega_0$ reads:

$$\dot{E} = \frac{\omega_0}{2\pi}\int_0^{2\pi/\omega_0} \dot{\mathcal{E}}(t)\mathrm{d}t = \frac{\omega_0}{2\pi}\int_0^{2\pi/\omega_0} m\dot{x}(t)f(t)\mathrm{d}t.$$

The average over the initial positions and velocities gives:

$$\langle \dot{E} \rangle = \frac{m}{T_0} \int_0^{T_0} \langle \dot{x}(t) f(t) \rangle \, dt, \tag{9.15}$$

where $T_0 = 2\pi/\omega_0$ is the oscillation period in the absence of the perturbations (i.e. $f(t) = 0$).

(5) The quantities of interest are directly related to the expression of $X(t)$:

$$x_0(t) = x(0)\cos(\omega_0 t) + \frac{\dot{x}(0)}{\omega_0}\sin(\omega_0 t)$$

$$\dot{x}_0(t) = -\omega_0 x(0)\sin(\omega_0 t) + \dot{x}(0)\cos(\omega_0 t)$$

$$x_1(t) = \frac{1}{\omega_0} \int_0^t d\tau \sin[\omega_0(t-\tau)] f(\tau). \tag{9.16}$$

$$\dot{x}_1(t) = \int_0^t d\tau \cos[\omega_0(t-\tau)] f(\tau). \tag{9.17}$$

(6) We just need to make the relation (9.15) explicit:

$$\langle \dot{x}(t) f(t) \rangle = \langle \omega_0^2 a(t)\dot{x}_0(t) + \omega_0^2 a(t)\dot{x}_1(t) \rangle$$

$$= \langle \omega_0^2 a(t)\dot{x}_1(t) \rangle = \omega_0^4 a(t) \int_0^t d\tau a(\tau)\cos[\omega_0(t-\tau)].$$

Indeed, the term $\langle \omega_0^2 a(t)\dot{x}_0(t) \rangle$ involves vanishing averages over the initial positions and velocities: $\langle x(0) \rangle = 0$ and $\langle \dot{x}(0) \rangle = 0$. We eventually obtain the desired relation $\langle \dot{E} \rangle = \Gamma$ with:

$$\Gamma = m\omega_0^4 \frac{1}{T_0} \int_0^{T_0} dt \int_0^t \cos(\omega_0 t') a(t) a(t-t') \, dt'.$$

This integral selects the frequency component ω_0 of the time correlation function $a(t)a(t-t')$ associated with the random displacement of the bottom of the trap. In the particular case $a(t) = a_0\cos(\omega_0 t)$, one finds the constant value[23] $\Gamma = (\pi/4)(m\omega_0^3 a_0^2)$. A linear increase of the mean total energy of the gas is therefore expected from the frequency component at ω_0 of the random position of the bottom of the trap.

(7) In this case, $f(t) = \omega_0^2 \epsilon(t)(x_0(t) + x_1(t))$, and

$$\langle \dot{E} \rangle = m\omega_0^2 \frac{1}{T_0} \int_0^{T_0} \langle G(t) \rangle \epsilon(t) dt, \tag{9.18}$$

to the lowest order. Indeed, the term in $\dot{x}_0(t)x_0(t)$ gives a vanishing contribution since $\langle \dot{x}^2(0) \rangle = \omega_0^2 \langle x^2(0) \rangle$. Using Eqs (9.16) and (9.17), we deduce directly the expression for $G(t)$:

$$G(t) = \int_0^t d\tau f(\tau) \left(x(0)\cos[\omega_0(2t-\tau)] + \frac{\dot{x}(0)}{\omega_0}\sin[\omega_0(2t-\tau)] \right).$$

---

[23] For $a(t) = a_0\cos(2\omega_0 t)$, i.e. for an oscillation of the position of the bottom of the trap at a frequency twice as large as the trap frequency, a simple calculation yields $\Gamma = 0$.

Taking into account only the lowest-order contribution of $f(\tau)$, i.e. that of $x_0(t)$, we have:

$$\frac{\langle G(t)\rangle}{m\omega_0^2} = \int_0^t \epsilon(\tau)\left\{ \langle x^2(0)\rangle \cos[2\omega_0(t-\tau)] + \frac{1}{\omega_0^2}\langle \dot{x}^2(0)\rangle \sin[2\omega_0(t-\tau)] \right\} d\tau.$$

Let us recall that the mean total energy $\langle E\rangle$ is, by definition, equal to:

$$\langle E\rangle = \left\langle \frac{1}{2}m\omega_0^2 x^2(0) + \frac{1}{2}m\dot{x}^2(0) \right\rangle.$$

Using the fact that the total energy of an harmonic oscillator is equally shared, on average, between the kinetic and the potentiel energy terms (this a consequence of the virial theorem studied in Problem 6.1), we can therefore write:

$$\langle x^2(0)\rangle = \frac{E}{m\omega_0^2} \qquad \text{and} \qquad \langle \dot{x}^2(0)\rangle = \frac{E}{m}.$$

Relation (9.18) can thus be rewritten as:

$$\frac{d}{dt}\langle E\rangle = \Gamma'\langle E\rangle \qquad \text{with}$$

$$\Gamma' = \frac{\sqrt{2}\omega_0^2}{T_0}\int_0^{T_0} dt \int_0^t \epsilon(t)\epsilon(t-t')\sin[2\omega_0 t' + \pi/4]\, dt'. \tag{9.19}$$

Let us consider the following example $\epsilon(t) = \epsilon_0 \cos(2\omega_0 t)$. In this case, one finds:

$$\Gamma' = \omega_0 \epsilon_0^2 \left(\frac{\pi}{4} - \frac{1}{16}\right).$$

In its most general form, Eq. (9.19) shows that Γ', and thus the heating rate, is sensitive the frequency component at $2\omega_0$ of the time correlation function $\epsilon(t)\epsilon(t-t')$. The increase of the mean energy due to this effect yields an exponential increase of the temperature. This effect, related to the frequency $2\omega_0$ (twice as large as the trapping frequency) is nothing but *parametric amplification* (see also Problem 8.4).

At this stage, we have all the ingredients to understand the experimental observations mentioned in the text. Indeed, if the currents that generate the magnetic field have a noise with a time correlation having a significant component at the frequency ω_0, a heating with a linear increase of the temperature is observed. If it has a significant component at a frequency $2\omega_0$, one observes an exponentiel increase of the temperature. Finally, in the absence of such noises, there is no heating. Magnetic traps used to confine ultracold atoms can reach, with special care, heating rates as low as a few nK per second.

▶ **Further reading.** For more details about these heating mechanisms, the interested reader may consult the following articles: T.A. Savard, K.M. O'Hara, and J.E. Thomas, *Laser-noise-induced heating in far-off resonance optical traps*, Phys. Rev. A **56**, 1095(R) (1997), and M.E. Gehm, K.M. O'Hara, T.A. Savard, and J.E. Thomas, *Dynamics of noise-induced heating in atom traps*, Phys. Rev. A **58**, 3914 (1998).

9.6 Evaporative cooling *

We consider a monoatomic ideal gas, at temperature T, whose N particles, of mass m, are constrained to move *in two dimensions*, in a "box" of area $\mathcal{S}$.

(1) Recall why the probability $\mathrm{d}P_{v_x}$ for an atom to have its velocity component along x between v_x and $v_x + \mathrm{d}v_x$ reads:

$$\mathrm{d}P_{v_x} = A \exp\left(-\frac{mv_x^2}{2k_B T}\right) \mathrm{d}v_x,$$

where k_B is Boltzmann's constant. Calculate A (we recall[24] the value of the Gaussian integral: $\int_{-\infty}^{\infty} \exp(-x^2)\mathrm{d}x = \sqrt{\pi}$).

(2) The two components x and y of the velocity being independent, show that the probability $\mathrm{d}^2P(\boldsymbol{v})$ for the velocity of the atom to be $\boldsymbol{v} = (v_x, v_y)$ within $(\mathrm{d}v_x, \mathrm{d}v_y)$ is given by:

$$\mathrm{d}^2P(\boldsymbol{v}) = A^2 \exp\left(-\frac{m(v_x^2 + v_y^2)}{2k_B T}\right) \mathrm{d}v_x\, \mathrm{d}v_y.$$

(3) Deduce from the previous result that the probability $\mathrm{d}P_v$ for the magnitude of $\boldsymbol{v}$ to be v within $\mathrm{d}v$ reads:

$$\mathrm{d}P_v = 2\pi A^2 v \exp\left(-\frac{mv^2}{2k_B T}\right) \mathrm{d}v.$$

Hint: work in the two-dimensional *velocity space*, in analogy with polar coordinates in real space.

(4) By a technique not described here, one can remove (or "evaporate"), in a very short time, all particles having a velocity v fulfilling $mv^2/2 > \eta k_B T$, where $\eta > 0$ is called the *evaporation parameter*. The other atoms are not affected. The number of particles, initially equal to N, then becomes $N' < N$. The particles removed are said to have been evaporated. Calculate and plot N'/N as a function of η.

---

[24]See the reminder on page 179 for the proof.

(5) Calculate the energy E' of the N' remaining particles, as a function of N, k_B, T, and η.

(6) After a long enough time, the remaining particles have rethermalized. Calculate the new equilibrium temperature T' of the gas (one will use the equipartition theorem, which implies that $E' = N'k_B T'$). Compare T' and T and plot T'/T versus η. Show that the gas cools down. Give a physical interpretation of this cooling.

(7) In quantum statistical mechanics, one defines the *phase-space density*, noted ρ, by $\rho = N\lambda^2/S$, where λ is the thermal de Broglie wavelength, defined as:

$$\lambda = \frac{h}{\sqrt{2\pi m k_B T}} \,.$$

Here, h is Planck's constant. Calculate the relative variation ρ'/ρ of the phase-space density upon evaporation. Plot ρ'/ρ versus η, and show that $\rho' > \rho$. Comment.

Solution

(1) The probability of having an energy E, for a particle in contact with a bath at temperature T, is given by the Boltzmann distribution: $P \propto \exp(-E/k_B T)$ (see the reminder at the end of the solution). Here, we have only the kinetic energy $E = mv_x^2/2$, and thus:

$$dP_{v_x} = A \exp\left(-\frac{mv_x^2}{2k_B T}\right) dv_x \,,$$

where A is a normalization constant. Its value is readily calculated by writing:

$$\int_{-\infty}^{\infty} A \exp\left(-\frac{mv_x^2}{2k_B T}\right) dv_x = A\sqrt{\frac{2k_B T}{m}} \int_{-\infty}^{\infty} \exp(-u^2) \, du = 1 \,,$$

whence $A = \sqrt{m/(2\pi k_B T)}$.

(2) The two components x and y of the velocity being independent, the probability $d^2 P(\boldsymbol{v})$ is equal to the product of the probabilities corresponding to v_x and v_y:

$$d^2 P(\boldsymbol{v}) = A^2 \exp\left(-\frac{m(v_x^2 + v_y^2)}{2k_B T}\right) dv_x \, dv_y \,.$$

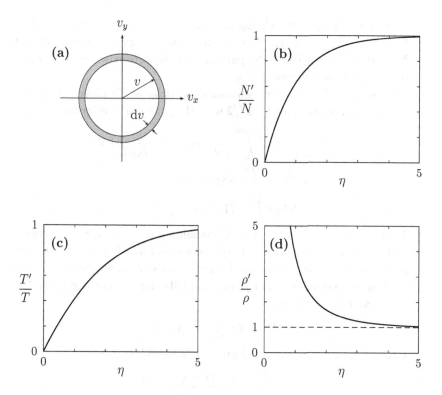

Fig. 9.12 (a) *Integration region in velocity space.* (b) N'/N *versus* η. (c) T'/T *versus* η. (d) ρ'/ρ *as a function of* η.

(3) We have to integrate the above probability over all values of v_x and v_y such that $\sqrt{v_x^2 + v_y^2}$ is between v and $v + \mathrm{d}v$, i.e., in the plane (v_x, v_y), the points located in the shaded area on Fig. 9.12(a). The area of this region is $2\pi v\,\mathrm{d}v$, we thus deduce:

$$\mathrm{d}P_v = 2\pi A^2 v \exp\left(-\frac{mv^2}{2k_B T}\right)\mathrm{d}v\,.$$

(4) The probability for a particle not to be evaporated is the same as that for having a velocity with a magnitude $v < v_0 = \sqrt{2\eta k_B T/m}$, therefore:

$$
\begin{aligned}
\frac{N'}{N} &= 2\pi A^2 \int_0^{v_0} v \exp\left(-\frac{mv^2}{2k_B T}\right)\mathrm{d}v \\
&= \int_0^{\eta} \exp\left(-\varepsilon\right)\mathrm{d}\varepsilon \\
&= 1 - e^{-\eta}\,,
\end{aligned}
$$

where we performed the change of variables $\varepsilon = mv^2/(2k_\mathrm{B}T)$. One does find that $N' \to N$ when $\eta \to \infty$ (no particle is evaporated) and that $N' \to 0$ when $\eta \to 0$ (all the particles are evaporated). Figure 9.12(b) shows N'/N as a function of η.

(5) The energy E' of the N' remaining particles is obtained simply by calculating the average of $mv^2/2$ with the probability distribution of v:

$$
\begin{aligned}
E' &= N 2\pi A^2 \int_0^{v_0} \frac{mv^2}{2}\, v \exp\left(-\frac{mv^2}{2k_\mathrm{B}T}\right) \mathrm{d}v \\
&= N k_\mathrm{B}T \int_0^{\eta} \varepsilon \exp(-\varepsilon)\, \mathrm{d}\varepsilon \\
&= N k_\mathrm{B}T \left[1 - (1+\eta)\mathrm{e}^{-\eta}\right].
\end{aligned}
$$

(6) After a long enough time, the N' remaining particles have rethermalized. The new temperature T' of the gas fulfills $E' = N'k_\mathrm{B}T'$ by the equipartition theorem: the total energy is shared among two quadratic degrees of freedom (v_x and v_y), each contributing with an energy $k_\mathrm{B}T/2$ per particle. We deduce:

$$
N'T' = NT \left[1 - (1+\eta)\mathrm{e}^{-\eta}\right],
$$

and, using the value obtained previously for N'/N, we get:

$$
\frac{T'}{T} = \frac{1 - (1+\eta)\mathrm{e}^{-\eta}}{1 - \mathrm{e}^{-\eta}}.
$$

Figure 9.12(c) shows T'/T as a function of η. One always has $T' \leqslant T$, there is therefore a cooling of the gas. Physically, the principle of this cooling mechanism can be understood easily: the most energetic particles of the distribution have been evaporated; the mean energy per remaining particle has thus decreased. The remaining particles therefore have, after rethermalization, a lower temperature.

(7) From the definition of phase-space density, $\rho \propto N/T$. The variation in phase-space density is thus:

$$
\frac{\rho'}{\rho} = \frac{N'/N}{T'/T} = \frac{1}{1 - (1+\eta)\mathrm{e}^{-\eta}}.
$$

It clearly appears that we always have $\rho' > \rho$. Figure 9.12(d) represents ρ'/ρ as a function of η.

One can repeat such an evaporation cycle many times in order to obtain large increases in phase-space density. Actually, one can also use a continuous scheme, in which the evaporation parameter η is kept constant by lowering the trap depth while the temperature decreases. In real experiments, the atoms are trapped in a three-dimensional harmonic trap, rather than in a two-dimensional box. This has an interesting consequence: one can show, indeed, that in that case, although the number of atoms decreases, the *spatial density* of the atomic cloud increases during the cooling, as well as the collision rate (the inverse of the mean time between binary collisions); this implies that the speed at which the evaporative cooling can take place (which is set by the thermalization rate, proportional to the collision rate) increases during the process. This phenomenon is called *runaway* evaporation.

This technique of *evaporative cooling*, proposed by Harald Hess to cool spin-polarized hydrogen[25], was demonstrated experimentally in the group of Daniel Kleppner and Thomas Greytak at MIT[26]. It allowed, in 1995, Eric Cornell, Carl Wieman, and Wolfgang Ketterle to observe a quantum phenomenon happening at very low temperatures, and called Bose–Einstein condensation, by cooling an alkali vapor to temperatures of hundreds of nK; for this achievement, they were awarded the Nobel prize in Physics in 2001.

▶ **Further reading.** The interested reader will find a similar model of evaporative cooling, but for more realistic experimental conditions (in particular, for gases trapped in three-dimensional harmonic or linear potentials) in the following article: K.B. Davis, M.-O. Mewes, and W. Ketterle, *An analytical model for evaporative cooling of atoms*, Appl. Phys. B **60**, 155 (1995). The Nobel lectures of Cornell, Wieman, and Ketterle are worth reading for a historical account of the achievement of Bose-Einstein condensation:

- E.A. Cornell and C.E. Wieman, *Bose–Einstein condensation in a dilute gas, the first 70 years and some recent experiments*, Rev. Mod. Phys. **74**, 875 (2002).

- W. Ketterle, *When atoms behave as waves: Bose–Einstein condensation and the atom laser*, Rev. Mod. Phys. **74**, 1131 (2002).

---

[25] H.F. Hess, *Evaporative cooling of magnetically trapped and compressed spin-polarized hydrogen*, Phys. Rev. B **34**, 3476 (1986).
[26] N. Masuhara, J.M. Doyle, J.C. Sandberg, D. Kleppner, T.J. Greytak, H.F. Hess, and G.P. Kochanski, *Evaporative cooling of spin-polarized atomic hydrogen*, Phys. Rev. Lett. **61**, 935 (1988).

Reminder: Boltzmann distribution

For a system at thermal equilibrium with a reservoir at temperature T, the probability of being in a particular state j is

$$p_j = \frac{1}{Z} \exp\left(-\frac{E_j}{k_{\mathrm{B}} T}\right), \tag{9.20}$$

where E_j is the energy of the state j of the system, and k_{B} is Boltzmann's constant. The normalization of this probability distribution is ensured by the prefactor $1/Z$, where Z, defined as

$$Z = \sum_{\text{all states } j} \exp\left(-\frac{E_j}{k_{\mathrm{B}} T}\right),$$

is called the *canonical partition function*. Note that a common mistake is to assume that the probability p_j given in Eq. (9.20) gives the probability of having an energy E_j: this is not the case! Indeed, the latter probability is given by multiplying p_j by the number of states having an energy E_j (i.e. the *degeneracy* of E_j).

Chapter 10

Celestial Mechanics

10.1 Closed trajectories in a central field *

We consider a planet of mass m evolving under the influence of a central force corresponding to the following potential:

$$V(r) = -\frac{K_1}{r} + \frac{K_2}{2r^2}.$$

Find the trajectories that correspond to closed orbits.

Solution

This exercise deals with a bound state, i.e. in which the distance between the two bodies never goes to infinity. The condition of having a *closed* orbit is stronger than the mere condition of having a bounded trajectory: a trajectory can obviously be bounded without being closed.

Since we deal with a central force, the angular momentum $L = r \times p$ is a conserved quantity. Indeed,

$$\frac{dL}{dt} = \frac{dr}{dt} \times p + r \times \frac{dp}{dt}.$$

The first term vanishes since $p = mv$ and $dr/dt = v$ are collinear, and the second term is also zero since the *central* force $F = dp/dt$ is collinear to r. The trajectory of the planet thus lies in a plane perpendicular to L. We use polar coordinates (r, θ) in this plane. The magnitude of the angular momentum reads $mr^2\dot{\theta}$. Let E be the total mechanical energy of the planet:

$$E = \frac{1}{2}m(\dot{r}^2 + r^2\dot{\theta}^2) + V(r) = \frac{1}{2}m\dot{r}^2 + V(r) + \frac{L^2}{2mr^2}.$$

213

The last equality is obtained by expressing $\dot{\theta}$ as a function of r using the conservation of angular momentum. The change of variable $u(\theta(t)) = 1/r(t)$ allows us to write the total energy as:

$$E = \frac{L^2}{2m}\left(\frac{du}{d\theta}\right)^2 + V(u) + \frac{L^2 u^2}{2m}. \qquad (10.1)$$

Indeed,

$$\dot{r} = \frac{dr}{dt} = \frac{d(1/u)}{dt} = \frac{d(1/u)}{du}\frac{du}{dt}$$

$$= -\frac{1}{u^2}\frac{du}{d\theta}\frac{d\theta}{dt} = -\frac{1}{u^2}\frac{du}{d\theta}\frac{Lu^2}{m} = -\frac{L}{m}\frac{du}{d\theta}.$$

An equation for the evolution of r as a function of the polar angle θ, or, equivalently, for the evolution of u as a function of θ, is obtained directly by differentiating the total energy (10.1) with respect to θ:

$$\frac{d^2 u}{d\theta^2} = -\frac{m}{L^2}\frac{d}{du}V_{\text{eff}}(u) \qquad (10.2)$$

with

$$V_{\text{eff}}(u) = V(u) + \frac{L^2 u^2}{2m} = -K_1 u + \frac{1}{2}K_2 u^2 + \frac{L^2}{2m}u^2.$$

Equation (10.2) is nothing but Newton's second law. Let us insist on the fact that, by using symmetries (or, equivalently, the existence of conserved quantities) we have reduced an *a priori* three-dimensional problem to a one-dimensional effective one.

It is convenient at this stage to introduce dimensionless quantities. We define $a = mK_1/L^2$ (the reciprocal of a length) and $\tilde{u} = u/a$. Equation (10.2) can then be rewritten as:

$$\frac{d^2\tilde{u}}{d\theta^2} = -\frac{d}{d\tilde{u}}\tilde{V}_{\text{eff}}(\tilde{u}) = -A\tilde{u} + 1, \qquad (10.3)$$

with

$$\tilde{V}_{\text{eff}}(\tilde{u}) = \frac{1}{2}A\tilde{u}^2 - \tilde{u} \qquad \text{and} \qquad A = \frac{mK_2}{L^2} + 1.$$

Equation (10.3) is the equation of motion of a harmonic oscillator; note however that here, it is the inverse of the radius r which oscillates sinusoidally as a function of the polar angle (and not as a function of time).

Figure 10.1 shows the effective potential $V_{\text{eff}}(u)$. One needs to consider two cases depending on the sign of the total energy. For $E < 0$ (energy E_1 in Fig. 10.1), the trajectory of the planet oscillates between the two radii $r_1 = 1/u_1$ and $r_2 = 1/u_2$. If $E > 0$ (energy E_2 in Fig. 10.1), the planet

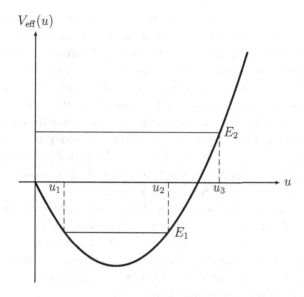

Fig. 10.1 *The effective potential V_{eff} has a parabolic shape. Depending on the sign of the energy, the trajectory corresponds to a bound state $(E < 0)$, or not $(E > 0)$.*

approaches $u = 0$ (it goes to infinity) and this takes an infinite time. A necessary condition to get a closed orbit is thus $E < 0$, but, as explained above, this is not a sufficient condition.

The solution of Eq. (10.3) is a sinusoidal motion with period $2\pi/\sqrt{A}$. In order to get a closed trajectory, it is thus necessary and sufficient that $\sqrt{A}$ is rational. Indeed, the trajectory is closed if and only if after an integer number of radial periods, the polar angle is a multiple of 2π. Let us stress the fact that this is a constraint on the angular momentum L of the planet, and therefore, *in fine*, on the initial conditions. This is to be contrasted with what happens in the case of a $1/r$ potential, for which *all* bound trajectories are closed.

10.2 Lagrange points ***

In this problem, we focus on the dynamics of a three-body system interacting *via* gravitational forces. This problem has been challenging generations of physicists since the resolution of the two-body problem by Newton himself. Although a general solution of this problem (under the form of series expansions) was found at the beginning of the twentieth century by Karl Sundman[1], a very interesting phenomenon was discovered as early as in 1772 by Joseph-Louis Lagrange: for the so-called *reduced* three-body problem (i.e. for the case in which one of the masses is negligible compared to the others, and thus does not disturb the motion of the two other bodies), there exist positions of *relative* equilibrium of the three bodies, some of them being stable (under some conditions). These equilibrium points are called Lagrange points, and they play a key role today in interplanetary navigation (e.g. for space probes). The goal of this problem is to prove this result in the simple case where the two bodies of finite mass orbit around their center of mass with a circular trajectory.

We thus consider two bodies A_1 and A_2, assumed to be point-like, with masses m_1 and m_2, interacting through gravitation. Let R be the distance separating them (R is constant since their orbit is assumed to be a circle). We define the inertial frame $\mathcal{R}_G$ by the center of mass O of the system and three "fixed stars". We note $r_1 = \overrightarrow{OA_1}$, $r_2 = \overrightarrow{OA_2}$, and $r = \overrightarrow{OM}$, where M stands for the point of negligible mass m.

Equilibrium points

(1) Calculate the angular velocity ω of the line $(A_1 A_2)$ in the frame $\mathcal{R}_G$.

(2) We now go to the rotating frame $\mathcal{R}'$ in which A_1 and A_2 are fixed. The x-axis is taken along the line $(A_1 A_2)$. Calculate, as a function of R and of $\alpha \equiv m_2/(m_1 + m_2)$, the coordinates A_1 and A_2 (that are now constant).

(3) Show, by going to the rotating frame, that finding the equilibrium points of mass $m \ll m_1, m_2$ consists in finding the stationary points of the potential:

$$E_{\mathrm{p}}(r) = -\frac{m\omega^2 r^2}{2} - \frac{Gmm_1}{\|r - r_1\|} - \frac{Gmm_2}{\|r - r_2\|},$$

where G is the gravitational constant. Where does the first term come from?

(4) Show that there are five equilibrium points L_i ($i = 1, \ldots, 5$) (all lying in the orbital plane of A_1, A_2), among which three lie on the axis $(A_1 A_2)$. For that purpose, one might sketch graphically $E_{\mathrm{p}}(r)$, after having introduced, first, adequate dimensionless parameters.

---

[1]As a result of a very slow convergence, this solution is in fact almost useless, and one can still today read in many texts that the three-body problem does not have any analytical solution.

(5) The three Lagrange points L_1 to L_3 are those located on the axis defined by the two massive bodies (L_1 lying between them, and L_2 close to the lighter one). In this question (and only this one) we assume that $m_1 \gg m_2$ (this is the case of the Earth–Sun system for instance). In this limit, show that, to the lowest order in α, the points L_1, L_2, L_3 have the following coordinates:

$$L_1 = R\left(1 - (\alpha/3)^{1/3}, 0\right),$$

$$L_2 = R\left(1 + (\alpha/3)^{1/3}, 0\right),$$

$$L_3 = -R\left(1 + 5\alpha/12, 0\right).$$

Calculate numerically these positions for the Earth–Sun system ($m_1 \simeq 2 \times 10^{30}$ kg and $m_2 \simeq 6 \times 10^{24}$ kg).

(6) We come back to the general case. Show that the Lagrange points L_4 and L_5 are such that the triangles $A_1 A_2 L_4$ and $A_1 A_2 L_5$ are equilateral. Hint: use polar coordinates with the origin in A_1.

Stability of the Lagrange points

In this section, we will show that the Lagrange points L_1 to L_3 are unstable, but that, under some conditions (to be established), the points L_4 and L_5 are stable equilibrium points.

(1) Explain why, in the rotating frame, it is not sufficient to check if the stationary points of E_p are minima or not in order to determine the stability of the equilibrium points. Why is it sufficient to study the stability in the plane (Oxy)?

(2) Show that by linearizing the equations of motion for small displacements $\varepsilon_x, \varepsilon_y$ around an equilibrium point L_i, we obtain:

$$\begin{cases} \ddot{\varepsilon}_x = 2\omega\dot{\varepsilon}_y - \varepsilon_x\,\partial_{xx}\phi - \varepsilon_y\,\partial_{xy}\phi, \\ \ddot{\varepsilon}_y = -2\omega\dot{\varepsilon}_x - \varepsilon_y\,\partial_{yy}\phi - \varepsilon_x\,\partial_{xy}\phi, \end{cases}$$

where $\phi = E_\mathrm{p}/m$ and where the partial derivatives are taken at the point L_i (the notation ∂_{xx} means $\partial^2/\partial x^2$). In order to simplify the calculations in the following questions, one will use dimensionless coordinates.

(3) Show that L_1, L_2 and L_3 are unstable (one will perform the calculations to lowest order in α). Give the characteristic time for the divergence of the trajectory.

(4) Show that L_4 and L_5 are stable, provided

$$\frac{m_1}{m_2} > \frac{25 + 3\sqrt{69}}{2}. \tag{10.4}$$

Is this condition fulfilled by the Earth–Sun system? And by the Earth–Moon system (we recall the mass of the Moon: 7.3×10^{22} kg)?

Solution

Equilibrium points

(1) We can directly apply Kepler's third law, which reads in the present case

$$\frac{T^2}{R^3} = \frac{4\pi^2}{G(m_1 + m_2)},$$

where $T = 2\pi/\omega$ is the orbital period. We thus get

$$\omega^2 = \frac{G(m_1 + m_2)}{R^3}. \tag{10.5}$$

(2) We define the basis (O, x, y, z) in the rotating frame by choosing the z-axis parallel to the angular velocity of the system, and the x-axis along the line $(A_1 A_2)$. The coordinates of A_1 and A_2 thus read, in the plane (x, y), $(-r_1, 0)$ and $(r_2, 0)$, with $r_1 + r_2 = R$. Moreover, by definition of the center of mass,

$$m_1 \boldsymbol{r}_1 + m_2 \boldsymbol{r}_2 = \boldsymbol{0}.$$

We deduce that $A_1 = (-\alpha R, 0)$ and $A_2 = (R(1 - \alpha), 0)$.

(3) When going to the rotating frame, we have to add, to the gravitational forces acting on the particle of mass m, inertial forces: the Coriolis force $-2m\boldsymbol{\omega} \times \boldsymbol{v}$ and the centrifugal force $m\omega^2 \overrightarrow{HM}$ (H being the projection of M onto the z-axis). In order to find equilibrium position, we do not need to take into account the Coriolis force (since, by definition, the relative velocity $\boldsymbol{v}$ is zero at equilibrium). The centrifugal force corresponds to the potential $-m\omega^2 (HM)^2/2$. Moreover, it is clear that the equilibrium points belong to the plane (Oxy), for otherwise the sum of gravitational forces would have a non-vanishing component along z, which could not be compensated for by the centrifugal force (always perpendicular to the rotation axis Oz). We thus need to look for the stationary points, in the plane (Oxy), of the potential:

$$E_{\mathrm{p}}(\boldsymbol{r}) = -\frac{m\omega^2 r^2}{2} - \frac{Gmm_1}{\|\boldsymbol{r} - \boldsymbol{r}_1\|} - \frac{Gmm_2}{\|\boldsymbol{r} - \boldsymbol{r}_2\|},$$

where the first term is the potential associated to the centrifugal force, and the two other terms the gravitational potential energy of M in the field of A_1 and A_2. It is useful, for the calculations to come, to use dimensionless quantities. Using R as the unit of length, and $Gm(m_1 + m_2)/R$ as the unit of energy, we have to find the stationary points of

$$\tilde{E}_{\mathrm{p}}(x,y) = -\frac{x^2 + y^2}{2} - \frac{1-\alpha}{\sqrt{(x+\alpha)^2 + y^2}} - \frac{\alpha}{\sqrt{(x-1+\alpha)^2 + y^2}},$$

where we made use of Eq. (10.5) giving ω, and where x and y are measured in units of R.

(4) It is easy to sketch the function $\tilde{E}_{\mathrm{p}}(x,y)$ (which is symmetric under the transformation $y \to -y$) whose stationary points need to be found [Fig. 10.2(a)]: the centrifugal term corresponds to an inverted paraboloid, and the gravitational potential digs into it two wells centered on A_1 and A_2. One easily understands that there are three stationary points lying on the axis $(A_1 A_2)$ [see the cut along Ox on Fig. 10.2(b)], and two away from it, symmetrically located with respect to (Ox). In the following questions, we will prove this assertion.

(5) The three Lagrange points L_1 to L_3 are located on the x-axis. Since $\tilde{E}_{\mathrm{p}}(x,y)$ is even in the variable y, all the points on the x-axis fulfill $\partial \tilde{E}_{\mathrm{p}}/\partial y = 0$; we thus need to find the values of x making $\partial \tilde{E}_{\mathrm{p}}/\partial x$ vanish. The function whose extrema must be found thus reads:

$$f(x) \equiv \tilde{E}_{\mathrm{p}}(x,0) = -\frac{x^2}{2} - \frac{1-\alpha}{|x+\alpha|} - \frac{\alpha}{|x+\alpha-1|}.$$

Let us measure distances from A_1 by introducing $u = x + \alpha$. We have

$$f(u) = -\frac{1}{2}(u-\alpha)^2 + \frac{\alpha-1}{|u|} - \frac{\alpha}{|u-1|}.$$

We first consider the case $0 < u < 1$ (Lagrange point L_1). In this range,

$$\frac{\mathrm{d}f}{\mathrm{d}u} = \alpha - u + \frac{1-\alpha}{u^2} - \frac{\alpha}{(u-1)^2}. \tag{10.6}$$

Solving $f'(u) = 0$ amounts to finding the roots of a polynomial of degree five, which cannot be done in closed form. We thus use a perturbative method, limiting ourselves to the lowest order in $\alpha \ll 1$. For $\alpha = 0$, the solution of Eq. (10.6) is simply $u = 1$. We thus look for a solution $u = 1 + \varepsilon$, with $|\varepsilon| \ll 1$. Keeping only the lowest-order terms in ε, we have:

$$\alpha - 1 - \varepsilon + (1-\alpha)(1-2\varepsilon) - \alpha/\varepsilon^2 = 0,$$

(a)

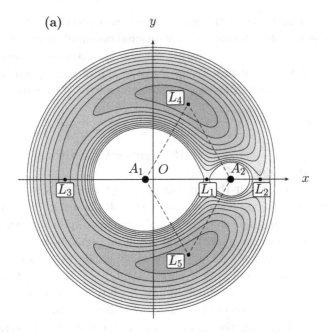

(b)

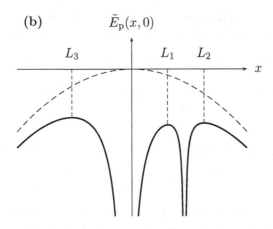

Fig. 10.2   (a) *Sketch of $\tilde{E}_p(x,y)$ for $m_1 = 10m_2$ (i.e. $\alpha = 1/11$). Darker regions correspond to higher potential energies. The surface $\tilde{E}_p(x,y)$ is obtained by "digging" two gravitational potential wells into an inverted paraboloid corresponding to the centrifugal term. The equilibrium points L_1 to L_5 correspond to stationary points of $\tilde{E}_p(x,y)$, i.e. to maxima (points L_4 and L_5) or saddle points (L_1, L_2 and L_3). (b) Plot of $\tilde{E}_p$ along the x-axis.*

or

$$\varepsilon = -\left(\frac{\alpha}{3}\right)^{1/3}.$$

Going back to dimensional coordinates, and noticing that α can safely be neglected in front of $\alpha^{1/3}$, we finally have

$$L_1 = R\left(1 - (\alpha/3)^{1/3}, 0\right).$$

We now consider L_2 ($u > 1$). By following exactly the same kind of argument, we obtain

$$\frac{df}{du} = \alpha - u + \frac{1-\alpha}{u^2} + \frac{\alpha}{(u-1)^2}.$$

whose approximate root is $u = 1 + \varepsilon$ with

$$\varepsilon = \left(\frac{\alpha}{3}\right)^{1/3},$$

which gives

$$L_2 = R\left(1 + (\alpha/3)^{1/3}, 0\right).$$

Finally, let us consider the case $u < 0$ (Lagrange point L_3). We must solve

$$\frac{df}{du} = \alpha - u + \frac{\alpha-1}{u^2} - \frac{\alpha}{(u-1)^2} = 0.$$

For $\alpha = 0$, the solution is $u = -1$; we thus look for a solution $u = -1 + \varepsilon$. We immediately find $\varepsilon = 7\alpha/12$, whence $x = -1 - 5\alpha/12$. The third Lagrange point

$$L_3 = -R\left(1 + 5\alpha/12, 0\right)$$

is therefore located slightly outside the orbit of A_2.
Numerically, for the Earth–Sun system, we find

$$\alpha = \frac{m_2}{m_1 + m_2} \simeq \frac{m_2}{m_1} \simeq \frac{6 \times 10^{24}}{2 \times 10^{30}} \simeq 3 \times 10^{-6}.$$

The points L_1 and L_2 are thus located approximately $R/100 \simeq 1.5$ million kilometers away from the Earth. The point L_3 is almost symmetrical of the Earth with respect to the Sun[2].

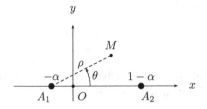

Fig. 10.3   *Polar coordinates (ρ, θ) with the origin in A_1.*

(6) In order to find the equilibrium points not located on the x-axis, let us introduce the polar coordinates (ρ, θ) with the origin in A_2. We then have (see Fig. 10.3)

$$\begin{cases} x = \rho \cos \theta - \alpha, \\ y = \rho \sin \theta. \end{cases}$$

The potential then reads, with these new variables:

$$\tilde{E}_{\mathrm{p}}(\rho, \theta) = -\frac{1}{2}\left(\rho^2 + \alpha^2 - 2\rho\alpha \cos\theta\right) + \frac{\alpha - 1}{\rho} - \frac{\alpha}{\left(\rho^2 - 2\rho\cos\theta + 1\right)^{1/2}}.$$

In order to find the stationary points, we write $\partial\tilde{E}_{\mathrm{p}}/\partial\rho = 0$ and $\partial\tilde{E}_{\mathrm{p}}/\partial\theta = 0$, giving

$$\begin{cases} -\rho + \alpha\cos\theta + \dfrac{1-\alpha}{\rho^2} + \dfrac{\rho\alpha - \alpha\cos\theta}{\mathcal{D}^{3/2}} = 0, \\[2mm] -\rho\alpha\sin\theta + \dfrac{\alpha\rho\sin\theta}{\mathcal{D}^{3/2}} = 0, \end{cases}$$

where $\mathcal{D} = \rho^2 - 2\rho\cos\theta + 1$. We are not interested here in the equilibrium points lying on the x-axis; we thus have $\sin\theta \neq 0$. From the second equation we then get $\mathcal{D} = 1$, whence:

$$\begin{cases} \rho^2 - 2\rho\cos\theta = 0, \\[1mm] -\rho + \alpha\cos\theta + \dfrac{1-\alpha}{\rho^2} + \alpha(\rho - \cos\theta) = 0, \end{cases}$$

which finally implies, since $\alpha \neq 1$:

$$\begin{cases} \rho = 1, \\ \cos\theta = 1/2. \end{cases}$$

We deduce that $\theta = \pm\pi/3$, and therefore there exist two equilibrium points L_4 and L_5 outside (Ox), forming equilateral triangles $A_1 A_2 L_4$ and $A_1 A_2 L_5$ with the two massive bodies A_1 and A_2 (independently of the values of m_1 and m_2).

---

[2] And thus is never visible from the Earth ... which has lead some sci-fi authors to make the L_3 point home of an unknown planet inhabited by dangerous aliens! However, as we shall see below, the point L_3 is unstable, ruling out this interesting possibility.

Stability of the Lagrange points

(1) The rotating frame $\mathcal{R}'$ in which we study the motion is *not inertial*. The Coriolis force thus plays a role in the equations of motion. Since it is not a conservative force, we cannot simply study the motion of the particle in a potential, and, in order to determine the stability of the equilibrium points, we must solve the equations of motions for small displacements around the equilibrium positions. In fact, the Coriolis force can stabilize equilibrium points that would be unstable otherwise (for instance maxima or saddle points, see Problems 4.1 and 4.2 for simple examples of such a *dynamical stabilization*). Here, we shall see that the points L_4 and L_5, although corresponding to maxima of the potential $\tilde{E}_p$, are actually stable.

First, let us note that all five Lagrange points correspond to a stable motion along z. Indeed, (i) the component of the Coriolis force $-2m\boldsymbol{\omega} \times \boldsymbol{v}$ is zero along this axis, since $\boldsymbol{\omega}//(Oz)$, and (ii) for a point located in $z \neq 0$, the sum of the gravitational forces is obviously oriented towards the (Oxy) plane: we thus have a restoring force along the z-axis. In the two following questions, we thus limit ourselves to the study of the stability in the plane (Oxy).

(2) We write the equations of motion for (x, y) in the neighborhood of the Lagrange point $L_i(x_i, y_i)$. By introducing $x = x_i + \varepsilon_x$ and $y = y_i + \varepsilon_y$ (with $|\varepsilon_x| \ll R$ and $|\varepsilon_y| \ll R$), Newton's second law

$$\ddot{\boldsymbol{r}} = -2\boldsymbol{\omega} \times \dot{\boldsymbol{r}} - \frac{1}{m}\boldsymbol{\nabla}E_p(\boldsymbol{r})$$

can be rewritten as:

$$\begin{cases} \ddot{\varepsilon}_x = 2\omega\dot{\varepsilon}_y - \dfrac{\partial}{\partial\varepsilon_x}\phi\left(x_i + \varepsilon_x, y_i + \varepsilon_y\right), \\ \ddot{\varepsilon}_y = -2\omega\dot{\varepsilon}_x - \dfrac{\partial}{\partial\varepsilon_y}\phi\left(x_i + \varepsilon_x, y_i + \varepsilon_y\right). \end{cases}$$

Since ε_x and ε_y are small quantities, we can expand the derivatives of ϕ to first order in $\varepsilon_x, \varepsilon_y$. Zeroth order terms vanish (because the point L_i is an equilibrium point, and thus a stationary point of E_p), and we obtain:

$$\begin{cases} \ddot{\varepsilon}_x = 2\omega\dot{\varepsilon}_y - \varepsilon_x\,\partial_{xx}\phi - \varepsilon_y\,\partial_{xy}\phi, \\ \ddot{\varepsilon}_y = -2\omega\dot{\varepsilon}_x - \varepsilon_y\,\partial_{yy}\phi - \varepsilon_x\,\partial_{xy}\phi, \end{cases}$$

We obtain dimensionless equations by using R as the unit of length, ω^{-1} as the unit of time, and $Gm(m_1 + m_2)/R$ as the unit of energy.

This amounts to performing the transformations $\omega \to 1$ and $\phi \to \tilde{E}_{\mathrm{p}}$ in the equations above, and yields:

$$\begin{cases} \ddot{\varepsilon}_x = 2\dot{\varepsilon}_y - \varepsilon_x \partial_{xx}\tilde{E}_{\mathrm{p}} - \varepsilon_y \partial_{xy}\tilde{E}_{\mathrm{p}}, \\ \ddot{\varepsilon}_y = -2\dot{\varepsilon}_x - \varepsilon_y \partial_{yy}\tilde{E}_{\mathrm{p}} - \varepsilon_x \partial_{xy}\tilde{E}_{\mathrm{p}}. \end{cases} \tag{10.7}$$

(3) We have to calculate the partial derivatives of second order of $\tilde{E}_{\mathrm{p}}$ in the various points L_1 to L_3. The first order derivatives are easy to get:

$$\partial_x \tilde{E}_{\mathrm{p}} = -x + \frac{(1-\alpha)(x+\alpha)}{[(x+\alpha)^2 + y^2]^{3/2}} + \frac{\alpha(x-1+\alpha)}{[(x-1+\alpha)^2 + y^2]^{3/2}},$$

$$\partial_y \tilde{E}_{\mathrm{p}} = -y + \frac{(1-\alpha)y}{[(x+\alpha)^2 + y^2]^{3/2}} + \frac{\alpha y}{[(x-1+\alpha)^2 + y^2]^{3/2}},$$

and from them we calculate:

$$\partial_{xx}\tilde{E}_{\mathrm{p}} = -1 + \frac{(1-\alpha)}{[(x+\alpha)^2 + y^2]^{3/2}} - 3\frac{(1-\alpha)(x+\alpha)^2}{[(x+\alpha)^2 + y^2]^{5/2}}$$
$$+ \frac{\alpha}{[(x-1+\alpha)^2 + y^2]^{3/2}} - 3\frac{\alpha(x-1+\alpha)^2}{[(x-1+\alpha)^2 + y^2]^{5/2}},$$

$$\partial_{yy}\tilde{E}_{\mathrm{p}} = -1 + \frac{(1-\alpha)}{[(x+\alpha)^2 + y^2]^{3/2}} - 3\frac{(1-\alpha)y^2}{[(x+\alpha)^2 + y^2]^{5/2}}$$
$$+ \frac{\alpha}{[(x-1+\alpha)^2 + y^2]^{3/2}} - 3\frac{\alpha y^2}{[(x-1+\alpha)^2 + y^2]^{5/2}},$$

$$\partial_{xy}\tilde{E}_{\mathrm{p}} = -3\frac{(1-\alpha)(x+\alpha)y}{[(x+\alpha)^2 + y^2]^{5/2}} - 3\frac{\alpha y(x-1+\alpha)}{[(x-1+\alpha)^2 + y^2]^{5/2}}.$$

We now just need to calculate explicitly the values of these three derivatives in $L_1(x_1, 0)$, $L_2(x_2, 0)$, and $L_3(x_3, 0)$, at lowest order in α, for $x_1 = 1 - (\alpha/3)^{1/3}$, $x_2 = 1 + (\alpha/3)^{1/3}$ and $x_3 = -1 - 5\alpha/12$. The result is shown in the following table:

	$\partial_{xx}\tilde{E}_{\mathrm{p}}$	$\partial_{yy}\tilde{E}_{\mathrm{p}}$	$\partial_{xy}\tilde{E}_{\mathrm{p}}$
L_1	-9	3	0
L_2	-9	3	0
L_3	-3	$7\alpha/8$	0

Case of L_1 and L_2. Eqs (10.7) read:

$$\begin{cases} \ddot{\varepsilon}_x = 2\dot{\varepsilon}_y + 9\varepsilon_x, \\ \ddot{\varepsilon}_y = -2\dot{\varepsilon}_x - 3\varepsilon_y. \end{cases}$$

We look for solutions of the form $\varepsilon_{x,y} = A_{x,y}e^{rt}$, where the exponent r is to be determined. The equations above imply the following set of equations for A_x, A_y:

$$\begin{cases} r^2 A_x = 2r A_y + 9A_x, \\ r^2 A_y = -2r A_x - 3A_y. \end{cases}$$

This set of equations has non-trivial solutions $(A_x, A_y) \neq (0,0)$ if and only if its determinant vanishes:

$$\begin{vmatrix} r^2 - 9 & -2r \\ 2r & r^2 + 3 \end{vmatrix} = 0.$$

The exponent r fulfills the biquadratic equation

$$r^4 - 2r^2 + 27 = 0,$$

whose solutions are:

$$r^2 = 1 + 2\sqrt{7} > 0 \quad \text{and} \quad r^2 = 1 - 2\sqrt{7} < 0.$$

We thus obtain as a possible value for r the positive one $r_1 = \left[1 + 2\sqrt{7}\right]^{1/2} \simeq 2.5$ corresponding to an exponentially diverging trajectory (with a time constant $\tau = 1/(r_1\omega)$, or, for the Earth–Sun system, 23 days). The Lagrange points L_1 and L_2 are therefore *unstable*.

Case of L_3. By reasoning exactly in the same way, we obtain the equation

$$\begin{vmatrix} r^2 - 3 & -2r \\ 2r & r^2 + 7\alpha/8 \end{vmatrix} = 0.$$

To the lowest non-vanishing order in α, the solutions of this biquadratic equation read:

$$r^2 = -1 \quad \text{and} \quad r^2 = \frac{21\alpha}{8}.$$

The first one gives oscillating solutions (since r is purely imaginary in that case), but the second shows that the motion is unstable, with an exponential divergence of the trajectory having a time constant $\tau = \sqrt{8/(21\alpha)}/\omega$ (about 57 years in the case of the Earth–Sun system). The Lagrange point L_3 is thus also unstable.

(4) For symmetry reasons, L_4 and L_5 have the same stability properties; we will thus study only the case of L_4, whose coordinates read $(1/2 - \alpha, \sqrt{3}/2)$. It is easy to calculate that at this point

$$\partial_{xx}\tilde{E}_{\mathrm{p}} = -\frac{3}{4}, \qquad \partial_{yy}\tilde{E}_{\mathrm{p}} = -\frac{9}{4}, \qquad \partial_{xy}\tilde{E}_{\mathrm{p}} = \frac{3\sqrt{3}}{4}(2\alpha - 1).$$

The same reasoning as above shows that we have to solve:

$$\begin{vmatrix} r^2 - 3/4 & -2r + 3\sqrt{3}(2\alpha - 1)/4 \\ 2r + 3\sqrt{3}(2\alpha - 1)/4 & r^2 - 9/4 \end{vmatrix} = 0,$$

i.e. the biquadratic equation:

$$r^4 + r^2 + \frac{27}{4}\alpha(1 - \alpha) = 0.$$

The two solutions for r^2 read

$$r^2 = \frac{-1 \pm \sqrt{1 - 27\alpha(1 - \alpha)}}{2}.$$

For the point L_4 to be stable, the exponent r must be purely imaginary, and thus r^2 must be a negative real number. This occurs if and only if:

$$0 < 1 - 27\alpha + 27\alpha^2 < 1$$

(the first inequality constrains r^2 to be real, while the second one imposes $r^2 < 0$). But, since by definition $0 < \alpha < 1$, one has always $1 - 27\alpha + 27\alpha^2 = 1 - 27\alpha(1 - \alpha) < 1$. The stability condition thus reads:

$$\alpha^2 - \alpha + \frac{1}{27} > 0.$$

Without loss of generality, we can assume that $m_1 \geqslant m_2$ (i.e. $\alpha \leqslant 1/2$), and *in fine* we get the condition

$$\frac{m_1}{m_2} > \frac{25 + 3\sqrt{69}}{2} \simeq 24.96.$$

This condition is by far fulfilled by the Earth–Sun system ($m_1/m_2 \simeq 3.3 \times 10^5$); but it is also fulfilled in the case of the Earth–Moon system ($m_1/m_2 \simeq 82$).

♦ **Remarks.**

(i) It is possible to show that even in the case where the mass m_3 of the third body is not negligible as compared to m_1 and m_2, the equilateral triangle configuration is still a position of relative equilibrium for the three masses (which all have circular trajectories). The proof is rather simple (see for instance D.F. Styer, *Simple derivation of Lagrange's three-body equilibrium*, Am. J. Phys. **58**, 917 (1990)). The study of the stability is more delicate and shows that the condition on the three masses reads

$$(m_1 + m_2 + m_3)^2 \geqslant 27(m_1m_2 + m_2m_3 + m_1m_3),$$

which, in the limit $m_3 \ll m_1, m_2$ amounts to Eq. (10.4).

(ii) The Lagrange points L_1 and L_2 of the Earth–Sun system are used for scientific missions. For instance the probe called SoHO (Solar and Heliospheric Observatory), dedicated to the observation of the Sun, was located around the point L_1. Since this point is unstable, the trajectory of the probe must be corrected periodically. As compared to a usual satellite orbit around the Earth, the advantage of the point L_1 for the observation of the Sun is that these can be done without interruption. In the same way, the probe called WMAP (Wilkinson Microwave Anisotropy Probe), dedicated to the study of the cosmic microwave background radiation, is located at L_2, allowing for a continuous observation of deep space without any disturbance from the Sun nor from the Earth.

(iii) The Lagrange points L_4 and L_5 being stable, many natural objects (asteroids) are encountered around the Lagrange points L_4 and L_5 of the Sun–Jupiter system; they thus move together with Jupiter around the Sun (with a delay or an advance of 60° with respect to the giant planet). As far as the Earth–Sun system is concerned, an asteroid called Cruithne, discovered in the 1980s, has an unusual trajectory with a "horseshoe" shape, enclosing the Lagrange points L_4 and L_5, in a quasi-synchronous fashion with the Earth[3].

▶ **Further reading.** For an introduction to some modern applications of the Lagrange points in space navigation, one can read the following article: S.D. Ross, *The interplanetary transport network*, American Scientist **94**(3), 230 (2006).

---

[3]P.A. Wiegert, K.A. Innanen, and S. Mikkola, *An asteroidal companion to the Earth*, Nature **387**, 685 (1997).

Chapter 11

Relativistic Mechanics

11.1 On the other side of the Galaxy *

The goal of this exercise is to emphasize one of the non-intuitive conse-
quences of special relativity, namely time dilation. Assume that we want
to reach a planet located on the other side of the galaxy, i.e. at a distance
of about 60,000 light-years, in a direct way (a straight line). We recall the
value of the speed of light: $c \simeq 3.00 \times 10^8$ m $\cdot$ s^{-1}. The rocket used by
the crew speeds up with a constant acceleration, equal to the one due to
gravity, $g = 9.8$ m$\cdot$s^{-2}, during the first half of the trip, and then decelerates
symmetrically. By how much will the crew members age? What about the
members of their families, who have stayed on Earth?

Solution

We first use an inertial frame coinciding at the time of departure with the
position of the Earth (and in which the velocity of the Earth is always very
small as compared to c). The distance to travel is $2\Delta = 60{,}000 \times (3 \times 10^8) \times 365 \times 24 \times 3{,}600 = 5.67 \times 10^{20}$ m. The first acceleration phase allows
the crew to travel a distance Δ in a time t^* and, for symmetry reasons, the
second phase also corresponds to a distance Δ covered in a time t^*. Finally,
we will assume[1] that the rest mass m of the rocket stays constant during
the mission. The equation of motion then reads, in one dimension (along
the straight trajectory):

$$\frac{\mathrm{d}(\gamma v)}{\mathrm{d}t} = g, \qquad \text{with} \qquad \gamma = \frac{1}{\sqrt{1 - v^2/c^2}} \qquad \text{and} \qquad v = \frac{\mathrm{d}x}{\mathrm{d}t}.$$

---

[1] Even though this assumption is not at all realistic!

Upon integration one thus gets $\gamma v = gt$ (as $v = 0$ at $t = 0$), giving

$$v(t) = \frac{gt}{\sqrt{1 + g^2t^2/c^2}}. \tag{11.1}$$

Note that for times such that $gt \ll c$, we recover the result from Newtonian mechanics, i.e. $v \simeq gt$, while in the opposite case we have, as expected, $v \simeq c$. We deduce the equation fulfilled by the coordinate x as a function of t:

$$\frac{dx}{dt} = \frac{gt}{\sqrt{1 + g^2t^2/c^2}}.$$

Separating the variables and integrating we obtain:

$$\Delta = \int_0^\Delta dx = \int_0^{t^\star} \frac{gt\,dt}{\sqrt{1 + g^2t^2/c^2}} = \frac{c^2}{2g} \int_0^{u^\star} \frac{du}{\sqrt{1 + u}},$$

where we have performed in the last integral the change of variable $u = g^2t^2/c^2$. Finally, we have

$$t^\star = \frac{c}{g}\sqrt{\left(1 + \frac{g\Delta}{c^2}\right)^2 - 1}.$$

Numerically, we find $t^\star \simeq 30{,}001$ years. The total duration of the trip in the inertial frame is therefore $2t^\star \simeq 60{,}002$ years. This is the time elapsed in the inertial frame we have chosen. Since the Earth moves in this frame with a velocity much smaller than c, this is also the duration of the trip as seen by observers on the Earth. The time needed for the trip is obviously larger than 60,000 years because the rocket needs a certain amount of time to reach a velocity close to c: for instance, the time $\tilde{t}$ needed to reach $c/2$ is from (11.1) $\tilde{t} = 1.76 \times 10^7$ s, i.e. slightly more than 200 days.

For the crew, time is given by the *proper time*. We thus need to calculate the variation of proper time $\tau^\star$. By definition,

$$\tau^\star = \int_0^{t^\star} \frac{dt}{\gamma} = \int_0^{t^\star} \sqrt{1 - v^2/c^2}\,dt = \int_0^{t^\star} \frac{dt}{\sqrt{1 + g^2t^2/c^2}} = \frac{c}{g}\sinh^{-1}\left(\frac{gt^\star}{c}\right).$$

Numerically we find $\tau^\star = 10.7$ years. For the crew, the trip thus lasts a little bit more than 21 years. This exemplifies one of the most counterintuitive predictions of special relativity: while the round trip lasts only about forty years for the crew, more than 120,000 years have elapsed on Earth. Figure 11.1 shows the distance Δ traveled by the rocket accelerating at g as function of the proper time $\tau^\star$ of the crew.

The above calculation illustrates the famous *twin paradox* raised by Paul Langevin in 1911: special relativity tells us that the measurement of

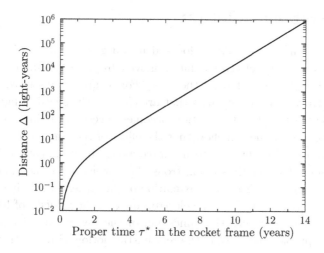

Fig. 11.1 *The distance Δ (in light-years) as a function of the proper time τ^* of the crew.*

a duration depends on the frame in which it is measured. If one brother of a twin pair stays on the Earth while his brother does the round-trip described above, the trip duration measured by the twin staying on the Earth is longer than the one measured by the traveling one. When they meet again, the traveling twin is thus younger than his brother. A contradiction seems to appear: one could argue that both twins acquire a large velocity compared to the other, and that the situation is symmetric! There is actually no contradiction, as the situation of both twins is not the same: the one staying on the Earth is always in an inertial frame, while his twin undergoes accelerations and decelerations which prevent him from staying in one and the same inertial frame. The above calculation clearly shows that the traveling brother will be younger than his brother when he comes back!

▶ **Further reading.** Nowadays, this effect of relativistic time dilation is not a purely academic matter: for instance, it must be properly accounted for in order to obtain accurate results with the Global Positioning System (GPS). A recent (and accessible) review of the subject can be found in N. Ashby, *Relativity in the Global Positioning System*, Living Rev. Relativity **6**, 1 (2003), available on the internet at the address: http://relativity.livingreviews.org/Articles/lrr-2003-1/.

11.2   Relativistic oscillator **

When an intense laser pulse is focused into a gas, the atoms in this gas are ionized and the electrons thus removed from the atoms oscillate in the electric field generated by the ions. For a large enough laser power, these electrons acquire velocities that are close to the speed of light and a relativistic treatment of the motion becomes necessary. Using the laws of electromagnetism, one can show that the effect of the ions on the electrons is well approximated by a restoring force along the propagation direction of the laser. The relativistic electrons then evolve in a one-dimensional potential $U(x)$. A realistic approximation of this potential is a linear one[2]: $U(x) = mc^2\eta|x|$, where m is the electron mass, c the velocity of light, and η a parameter which has the dimension of the inverse of a length. The goal of the problem is to study the relativistic motion of an electron in this one-dimensional linear trapping potential.

(1) Write down the equation of motion.
(2) Deduce from it the expression of a constant of motion.
(3) Calculate the oscillation period of a particle in this potential. Comment.
(4) Calculate explicitly the position $x(t)$ as a function of the velocity $\dot{x}$ in the phase space sector $x > 0$ and $\dot{x} > 0$.
(5) Plot the trajectories in phase space (in order to be able to easily compare the influence of relativistic effects, one may normalize the trajectories to the maximal position and velocity). Comment.

We recall that:

$$\int_0^x \frac{1 + au}{\sqrt{u}\sqrt{2 + au}}\, du = \sqrt{x(2 + ax)}. \qquad (11.2)$$

Solution

(1) The equation of motion in special relativity reads, as in the classical limit:

$$\frac{dp}{dt} = -\frac{dU}{dx}; \qquad (11.3)$$

---

[2]D. Teychenne, G. Bonnaud, and J.-L. Bobin, *Oscillatory relativistic motion of a particle in a power-law or sinusoidal-shaped potential well*, Phys. Rev. E **49**, 3253 (1994).

however, the expression for the momentum p is:

$$p = m\gamma v = m\gamma \dot{x} = m\gamma \frac{\mathrm{d}x}{\mathrm{d}t} = \frac{m\dot{x}}{\sqrt{1 - \dot{x}^2/c^2}}.$$

(2) A constant of motion is a function of the position and velocity which stays constant along the trajectory. In one dimension, both in relativistic and classical mechanics, the only possible integral of motion is the energy, since writing Newton's second law for a conservative force is equivalent to giving the expression of the total energy. To deduce the latter from the equation of motion, we multiply Eq. (11.3) by the velocity $\dot{x}$:

$$m\dot{x}\frac{\mathrm{d}}{\mathrm{d}t}\frac{\dot{x}}{\sqrt{1 - \dot{x}^2/c^2}} = -\frac{\mathrm{d}x}{\mathrm{d}t}\frac{\mathrm{d}U}{\mathrm{d}x} = -\frac{\mathrm{d}U}{\mathrm{d}t}. \tag{11.4}$$

The left hand side of Eq. (11.4) can be rewritten as

$$m\dot{x}\frac{\mathrm{d}}{\mathrm{d}t}\frac{\dot{x}}{\sqrt{1 - \dot{x}^2/c^2}} = \frac{m\dot{x}\ddot{x}}{(1 - \dot{x}^2/c^2)^{3/2}} = mc^2\frac{\mathrm{d}}{\mathrm{d}t}\frac{1}{\sqrt{1 - \dot{x}^2/c^2}} = \frac{\mathrm{d}}{\mathrm{d}t}(m\gamma c^2).$$

We finally recover the conservation of energy:

$$\frac{\mathrm{d}E}{\mathrm{d}t} = 0 \quad \text{with} \quad E = m\gamma c^2 + U(x).$$

Let us notice that the non-relativistic expression for the energy is recovered by expanding γ for velocities $v \ll c$:

$$\gamma = \left(1 - \frac{\dot{x}^2}{c^2}\right)^{-1/2} \simeq 1 + \frac{\dot{x}^2}{2c^2}.$$

In this limit, we find:

$$E \simeq mc^2 + \frac{1}{2}m\dot{x}^2 + U(x),$$

i.e. the expression of the energy as the sum of a kinetic energy term $(m\dot{x}^2/2)$ and of a potential energy term $U(x)$ (within the constant mc^2 that can be absorbed in the definition of E).

(3) The calculation of the period relies on energy conservation. The motion of the particle has an amplitude a. At the turning point $x = a$, the velocity vanishes. We thus deduce the relationship between γ and the position x:

$$E = mc^2\gamma + U(x) = mc^2 + U(a) \Longrightarrow \gamma = 1 + \frac{U(a) - U(x)}{mc^2}.$$

We therefore obtain an explicit relationship between $\dot{x}$ and x:

$$\frac{\dot{x}^2}{c^2} = \eta\frac{(a - |x|)(2 + \eta(a - |x|))}{[1 + \eta(a - |x|)]^2}. \tag{11.5}$$

Let us consider the phase space sector for which $x(t) > 0$ and $\dot{x}(t) > 0$. We choose the origin of time $t = 0$ when the particle is at $x = 0$ and moves towards $x > 0$. The velocity is then positive, and we deduce from Eq. (11.5), after separation of the variables:

$$\frac{1 + \eta(a - x)}{\sqrt{(a - x)[2 + \eta(a - x)]}}\, dx = c\sqrt{\eta}\, dt,$$

or, equivalently, using Eq. (11.2):

$$-d\left(\sqrt{(a - x)\left[2 + \eta(a - x)\right]}\right) = d\left(ct\sqrt{\eta}\right). \tag{11.6}$$

By definition, a quarter of a period $(T/4)$ has elapsed between the time when the particle is at $x = 0$ and the one when it is at the turning point $x = a$. From Eq. (11.6) we deduce the expression of the oscillation period:

$$T = 4\sqrt{\frac{a}{\eta c^2}}\sqrt{2 + \eta a}. \tag{11.7}$$

One easily checks that a non-relativistic particle in the same potential has the oscillation period $T_0 = 4\sqrt{2a/\eta c^2}$. The term ηa in Eq. (11.7) is responsible for relativistic corrections. In the ultra-relativistic regime $\eta a \gg 1$ (or $E \gg mc^2$), the oscillation period is proportional to the amplitude $T \propto a$, whereas in the classical regime, $T \simeq T_0 \propto \sqrt{a}$ (this is a well-known result, for instance in the case of free fall: the fall time increases as the square root of the height).

(4) The expression for $x(t)$ is obtained by integrating Eq. (11.6) from $x = 0$ to $x > 0$. We find:

$$x = a + \frac{1}{\eta}\left(1 - \sqrt{1 + \eta^2 c^2 (T/4 - t)^2}\right) \quad \text{and} \quad \dot{x} = \frac{\eta c^2 (T/4 - t)}{\sqrt{1 + \eta^2 c^2 (T/4 - t)^2}}.$$

(5) Figure 11.2 shows trajectories in phase space for increasing values of the parameter ηa. When ηa is large, the apparent mass $m\gamma$ of the particle is larger and the velocity decreases more slowly because of an increased inertia. We thus understand why the velocity "plateau" around the equilibrium position gets flatter and flatter when the particle energy increases to relativistic values.

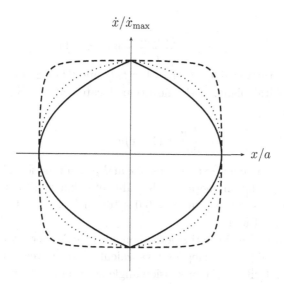

Fig. 11.2 *Trajectories in phase space for the classical regime $\eta a = 0.1$ (solid line), for the intermediate regime $\eta a = 1$ (dotted line), and for the relativistic regime $\eta a = 10$ (dashed line).*

11.3 Perihelion precession of Mercury **

In this problem, we study a historically important application of relativistic mechanics. The orbit of Mercury, the closest planet to the Sun, is not exactly a closed ellipse: the perihelion slowly precesses. It was shown by the astronomers of the nineteenth century that the influence of the other planets plays a role in this effect, but cannot account for the numerical value of the observed precession. Einstein's special relativity adds an extra contribution which has the right sign, but cannot explain the observations quantitatively. Only by using general relativity (which is beyond the scope of this book) can one reconcile theory and observations. Here, we establish the contribution of the effects of special relativity.

The Sun has a mass $M = 2.0 \times 10^{30}$ kg, and the mass of Mercury is $m = 3.3 \times 10^{23}$ kg. The gravitational constant is $G = 6.7 \times 10^{-11}$ m$^3 \cdot$ kg$^{-1} \cdot$ s^{-2}, and the speed of light will be taken equal to $c = 3.0 \times 10^8$ m $\cdot$ s^{-1}.

(1) Show that the trajectory of Mercury in the Sun gravitational field lies within a plane. One will denote by L the angular momentum.
(2) Show, using Newton's second law, that the total energy E of Mercury

reads:

$$E = -\frac{GMm}{r} + mc^2(\gamma - 1). \tag{11.8}$$

(3) Show that the variable $u[\theta(t)] = 1/r(t)$ (r and θ being polar coordinates in the orbital plane, with the origin located at the Sun) obeys the equation

$$\frac{d^2u}{d\theta^2} + (1 - \epsilon)u = \frac{1 - \epsilon}{p},$$

and give the explicit expressions of ϵ and p as a function of G, m, M, c and E. Calculate numerically the value of ϵ, knowing that astrometric measurements give a value $L = 9.0 \times 10^{38}$ m$^2 \cdot$ kg $\cdot$ s^{-1} for the angular momentum of Mercury.

(4) Deduce from the above the perihelion advance $\Delta\theta$ per revolution. Mercury's orbital period being 88 days, calculate the correction δ_{sr} given by special relativity to the precession angle per century. General relativity predicts, in good agreement with observations, $\delta_{gr} = 43''$. Comment.

Solution

(1) The gravitational interaction between two bodies of masses m and M gives rise to the central force

$$\mathbf{F} = -\frac{GMm}{r^2}\mathbf{u}_r.$$

The angular momentum $\mathbf{L} = \mathbf{r} \times \mathbf{p}$ is thus a constant of motion. Indeed one has

$$\frac{d\mathbf{L}}{dt} = \frac{d\mathbf{r}}{dt} \times \mathbf{p} + \mathbf{r} \times \frac{d\mathbf{p}}{dt} = \mathbf{0}.$$

The first term $(d\mathbf{r}/dt \times \mathbf{p})$ vanishes as $\mathbf{v} = d\mathbf{r}/dt$ is parallel to $\mathbf{p} = m\gamma\mathbf{v}$. The second term $(\mathbf{r} \times d\mathbf{p}/dt)$ is also zero according to Newton's second law $d\mathbf{p}/dt = \mathbf{F}$, since $\mathbf{F} \propto \mathbf{r}$ which implies that $d\mathbf{p}/dt$ is parallel to $\mathbf{r}$. Since the angular momentum $\mathbf{L}$ is a constant, the motion occurs within a plane. Indeed, if the trajectory went out of the plane normal to $\mathbf{L}$, the angular momentum would necessarily change at that time. We notice that the symmetry arising from the central character of the force yields a constraint on the trajectory, as *in fine* it reduces the dimension of the space in which it evolves from three to two. In the following, an extra constant of motion — the total energy — arises and the problem becomes effectively one-dimensional. This reduction of dimensionality due to symmetries (or, equivalently, to the existence of constants of motion) allows for a simple solution of an *a priori* complex problem.

(2) Newton's second law reads

$$\frac{\mathrm{d}\boldsymbol{p}}{\mathrm{d}t} = -\frac{GMm}{r^2}\boldsymbol{u}_r. \tag{11.9}$$

Since $\boldsymbol{p} = m\gamma\boldsymbol{v}$, we have to calculate the time derivative of γ; for that purpose it is convenient to use logarithmic differentiation:

$$\gamma = \frac{1}{\sqrt{1 - v^2/c^2}} \implies \ln\gamma = -\frac{1}{2}\ln\left(1 - \frac{v^2}{c^2}\right) \implies \frac{\mathrm{d}\gamma}{\gamma} = \frac{1}{c^2}\frac{\boldsymbol{v}\cdot\mathrm{d}\boldsymbol{v}}{1 - v^2/c^2}.$$

We finally have

$$\frac{\mathrm{d}\gamma}{\mathrm{d}t} = \frac{\gamma^3}{c^2}\boldsymbol{v}\cdot\frac{\mathrm{d}\boldsymbol{v}}{\mathrm{d}t}.$$

Energy conservation can be obtained, as in the non-relativistic case, by multiplying Newton's second law by the velocity $\boldsymbol{v}$. The l.h.s. of Eq. (11.9) reads:

$$\begin{aligned}
\boldsymbol{v}\cdot\frac{\mathrm{d}\boldsymbol{p}}{\mathrm{d}t} &= mv^2\frac{\mathrm{d}\gamma}{\mathrm{d}t} + m\gamma\boldsymbol{v}\cdot\frac{\mathrm{d}\boldsymbol{v}}{\mathrm{d}t} \\
&= m\gamma\boldsymbol{v}\cdot\frac{\mathrm{d}\boldsymbol{v}}{\mathrm{d}t}\left(1 + \frac{v^2}{c^2}\gamma^2\right) \\
&= m\gamma^3\boldsymbol{v}\cdot\frac{\mathrm{d}\boldsymbol{v}}{\mathrm{d}t} = \frac{\mathrm{d}}{\mathrm{d}t}(m\gamma c^2).
\end{aligned} \tag{11.10}$$

The force term allows us to write

$$\boldsymbol{v}\cdot\frac{\mathrm{d}\boldsymbol{p}}{\mathrm{d}t} = -\frac{GMm}{r^2}\boldsymbol{u}_r\cdot\boldsymbol{v} = -\frac{GMm}{r^2}\frac{\mathrm{d}r}{\mathrm{d}t} = \frac{\mathrm{d}}{\mathrm{d}t}\left(\frac{GMm}{r}\right). \tag{11.11}$$

By combining Eqs (11.9), (11.10), and (11.11), we finally establish the conservation of the total energy E:

$$\frac{\mathrm{d}}{\mathrm{d}t}\left(mc^2(\gamma - 1) - \frac{GMm}{r}\right) \equiv \frac{\mathrm{d}E}{\mathrm{d}t} = 0.$$

The energy being defined within an additive constant, we have subtracted mc^2, in such a way that a Taylor expansion of the kinetic energy term $mc^2(\gamma - 1)$ does give $mv^2/2$ in the limit $v \ll c$.

(3) The reasoning will be similar to that followed in the non-relativistic case. First, we use polar coordinates in the plane containing the trajectory. Then we perform the change of variables $u(\theta) = 1/r(t)$. Let us

express first the angular momentum in polar coordinates. The velocity can be rewritten as:

$$v = \frac{dr}{dt} = \frac{d(ru_r)}{dt} = \frac{dr}{dt}u_r + r\frac{du_r}{dt}$$

$$= \dot{r}u_r + r\frac{du_r}{d\theta}\frac{d\theta}{dt} = \dot{r}u_r + r\dot{\theta}u_\theta.$$

We deduce:

$$L = r \times p = m\gamma r u_r \times (\dot{r}u_r + r\dot{\theta}u_\theta) = m\gamma r^2 \dot{\theta} u_z.$$

This expression of the angular momentum can be used to write p as a function of u:

$$p = m\gamma(\dot{r}u_r + r\dot{\theta}u_\theta) = m\gamma r^2 \dot{\theta}\left(\frac{dt}{d\theta}\frac{dr}{dt}\frac{1}{r^2}u_r + \frac{1}{r}u_\theta\right)$$

$$= L\left(-\frac{du}{d\theta}u_r + uu_\theta\right).$$

The time derivative of the momentum p thus reads:

$$\frac{dp}{dt} = L\left[-\frac{d}{dt}\left(\frac{du}{d\theta}\right)u_r - \frac{du}{d\theta}\frac{du_r}{dt} + \frac{du}{dt}u_\theta + u\frac{du_\theta}{dt}\right]$$

$$= L\left[-\frac{d\theta}{dt}\frac{d}{d\theta}\left(\frac{du}{d\theta}\right)u_r - \frac{du}{d\theta}\frac{d\theta}{dt}\frac{du_r}{d\theta} + \frac{d\theta}{dt}\frac{du}{d\theta}u_\theta + u\frac{d\theta}{dt}\frac{du_\theta}{d\theta}\right]$$

$$= -L\dot{\theta}\left[\frac{d^2u}{d\theta^2} + u\right]u_r. \tag{11.12}$$

We have used the well-known identities $du_r/d\theta = u_\theta$ and $du_\theta/d\theta = -u_r$. Equations (11.9) and (11.12) allow us to write:

$$-L\dot{\theta}\left[\frac{d^2u}{d\theta^2} + u\right] = -\frac{GMm}{r^2},$$

or, equivalently,

$$\frac{d^2u}{d\theta^2} + u = \frac{GMm^2}{L^2}\gamma.$$

The expression of γ as a function of u is obtained from energy conservation (11.8). We finally obtain:

$$\frac{d^2u}{d\theta^2} + (1 - \epsilon)u = \frac{1 - \epsilon}{p}. \tag{11.13}$$

with

$$\epsilon = \left(\frac{GMm}{Lc}\right)^2 \quad \text{and} \quad \frac{1 - \epsilon}{p} = \frac{GMm^2}{L^2}\left(1 + \frac{E}{mc^2}\right).$$

The solution of Eq. (11.13) has the form:

$$u(\theta) = \frac{1 + e\cos[(\theta - \theta_0)\sqrt{1 - \epsilon}]}{p}.$$

With the numerical values of the text we obtain $\epsilon = 2.7 \times 10^{-8}$. This value is very small because the maximal velocity of Mercury is much smaller than c, and ϵ accounts for the small correction that relativistic effects induce on the non-relativistic results.

(4) When Mercury reaches its perihelion again, $(\theta - \theta_0)\sqrt{1 - \epsilon} = 2\pi$. The perihelion thus has an advance of $\Delta\theta = \epsilon\pi$ since $(\theta - \theta_0) = 2\pi/\sqrt{1 - \epsilon} \simeq 2\pi + \epsilon\pi$. With the numerical values given in the text, we find $\Delta\theta = 8.4 \times 10^{-8}$. This is equivalent to an advance per century, expressed in arc-seconds:

$$\delta_{\mathrm{sr}} = \frac{\Delta\theta \times 100 \times 365}{88} \times \frac{360 \times 3{,}600}{2\pi} = 7.2''.$$

Clearly, this result cannot account for the observed advance of $43''$ per century[3]. Confronted by this anomalous precession, the astronomers of the nineteenth century assumed wrongly that one or several unknown celestial bodies were orbiting between the Sun and Mercury's orbit, perturbing the motion of Mercury *via* the gravitational interaction. As Mercury is the planet closest to the Sun, it is located in the region where the Sun's gravitational field is the strongest, and thus it experiences the largest modifications with respect to Newton's laws that are brought by Einstein's general relativity.

11.4 Electron motion in a plane wave ***

In this exercise, we study the motion of an electron (mass m, charge q) subjected to an intense electromagnetic plane wave. When the wiggling energy of the electron becomes comparable to its rest energy mc^2, the magnetic component of the Lorentz force considerably modifies the trajectory of the electron. This regime corresponds to intensities (power per unit surface) in the range of a few 10^{18} W·cm^{-2}. Such intensities can be achieved by compressing intense light pulses on very short durations.

---

[3]In fact, the total advance of the perihelion of Mercury is $574''$ per century. Taking into account, within the framework of classical mechanics, the perturbations of Mercury's trajectory due to the attraction of other planets allows us to explain a precession of $531''$ per century. Only general relativity can fully explain the missing $43''$.

We assume that the incident monochromatic plane wave is applied at $t = 0$. In the following, the duration of the pulse is supposed to be infinite. We introduce the phase term $\phi(z,t) = kz - \omega t$, and recall the definition of the proper time τ in special relativity:

$$\frac{dt}{d\tau} = \gamma = \frac{1}{\sqrt{1 - v^2/c^2}},$$

where c is the speed of light in vacuum. The dispersion relation of light waves in vacuum is $\omega = kc$. The vector potential $\boldsymbol{A}$ of the transverse plane wave experienced by the electron reads:

$$\boldsymbol{A}(z,t) = \boldsymbol{A}_\perp(z,t) = \begin{pmatrix} a_x \cos\phi(z,t) \\ a_y \sin\phi(z,t) \\ 0 \end{pmatrix},$$

where the index $\perp$ refers to the coordinates x and y, transverse to the propagation direction z. We assume that the amplitudes a_x and a_y are constant and that the electron is initially (at $t = 0$) at the origin of the laboratory frame $\boldsymbol{r}_0 = \boldsymbol{0}$ with a longitudinal velocity v_0.

(1) Give the expression of the electric field $\boldsymbol{E}$ and of the magnetic field $\boldsymbol{B}$ as a function of ϕ, ω, k, a_x and a_y.

(2) Write down the equations of motion of the electron in the field of the plane wave.

(3) Deduce that the quantities $\boldsymbol{\kappa}_1 = \boldsymbol{p}_\perp + q\boldsymbol{A}_\perp$ and $\Omega = k\gamma(v_z - c)$ are constants of motion.
For the sake of simplicity, we assume in the following that the initial conditions are such that $\boldsymbol{\kappa}_1 = \boldsymbol{0}$.

(4) Show that

$$\phi(z(t), t) = \Omega\tau.$$

(5) Solve the equation of motion for the particular case of a circularly polarized plane wave. Give the explicit expression for $x(\tau)$, $y(\tau)$, and $z(\tau)$ (hint: show that, under these assumptions, γ is constant).

(6) Solve the equation of motion for the particular case of a linearly polarized plane wave ($a_x = 0$). Give the explicit expression for $y(\tau)$ and $z(\tau)$. Sketch the trajectory of the electron for $v_0 = 0$. Comment.

Solution

In this exercise we work out the analytic solution of the motion of a relativistic electron in a plane wave. It constitutes a good practice in the manipulation of partial and total derivatives. The amplitude of the electric and magnetic field are assumed to be constant. The reader interested by the effects that arise when these amplitudes depend on transverse coordinates can study Problem 8.5, entitled "Ponderomotive force", where the non-relativistic case is investigated in details.

(1) We choose the Coulomb gauge. The expression for the electric field is obtained by differentiating the vector potential with respect to time:

$$E = -\frac{\partial A}{\partial t} = -\frac{\partial \phi}{\partial t}\frac{\mathrm{d}A}{\mathrm{d}\phi} = \omega \begin{pmatrix} -a_x \sin\phi \\ a_y \cos\phi \\ 0 \end{pmatrix}.$$

In the relativistic domain, the contribution of the magnetic field to the dynamics of the electron can no longer be neglected. Its expression is obtained from the vector potential by:

$$B = \nabla \times A = \begin{pmatrix} -\partial_z A_y \\ \partial_z A_x \\ \partial_x A_y - \partial_y A_x \end{pmatrix} = k \begin{pmatrix} -a_y \cos\phi \\ -a_x \sin\phi \\ 0 \end{pmatrix}.$$

(2) The equation of motion reads:

$$\frac{\mathrm{d}p}{\mathrm{d}t} = qE + qv \times B. \tag{11.14}$$

Let us calculate the cross product

$$v \times B = \begin{pmatrix} -\dot{z}\partial_z A_x \\ -\dot{z}\partial_z A_y \\ \dot{x}\partial_z A_x + \dot{y}\partial_z A_y \end{pmatrix}. \tag{11.15}$$

Indeed, the components of the vector potential depend only on the longitudinal variable z. For non-relativistic velocities $v \ll c$, the magnetic contribution of the Lorentz force is negligible as compared to the electric one: $\|qv \times B\| \ll \|qE\|$, since $B \sim E/c$. Here, we consider the relativistic regime and we have to take into account its contribution. As a result, the Lorentz force acquires a longitudinal component. In the following, we discuss the conditions on the polarization of the plane wave under which the magnetic component of the Lorentz force yields a clear change on the relativistic trajectory as compared to its non-relativistic counterpart.

(3) Taking Eq. (11.16) into account, the equation of motion (11.14) can be recast as

$$\frac{\mathrm{d}p_\perp}{\mathrm{d}t} = -q\frac{\partial A_\perp}{\partial t} - q\dot{z}\frac{\partial A_\perp}{\partial z} = -q\frac{\mathrm{d}A_\perp}{\mathrm{d}t},$$

whence

$$\frac{\mathrm{d}}{\mathrm{d}t}(\boldsymbol{p}_\perp + q\boldsymbol{A}_\perp) = \boldsymbol{0}.$$

We deduce that the quantity $\boldsymbol{\kappa}_1 = \boldsymbol{p}_\perp + q\boldsymbol{A}_\perp$ is a constant of motion. Equation (11.14) projected on the z axis yields

$$\frac{\mathrm{d}p_z}{\mathrm{d}t} = q\dot{x}\partial_z A_x + \dot{y}\partial_z A_y = q\boldsymbol{v}\cdot\frac{\partial\boldsymbol{A}}{\partial z}.$$

To determine the other constant of motion we use an approach that is reminiscent of the one used in non-relativistic mechanics to obtain the conservation of the total energy from Newton's second law. We take the dot product of the equation of motion with the velocity:

$$m\boldsymbol{v}\cdot\frac{\mathrm{d}\boldsymbol{p}}{\mathrm{d}t} = mq\boldsymbol{v}\cdot\boldsymbol{E} = -mq\boldsymbol{v}\cdot\frac{\partial\boldsymbol{A}}{\partial t} = mqc\boldsymbol{v}\cdot\frac{\partial\boldsymbol{A}}{\partial z} = mc\frac{\mathrm{d}p_z}{\mathrm{d}t}. \quad (11.16)$$

The equality relies on the fact that the vector potential depends on time only through the phase variable ϕ, which implies:

$$\frac{\partial\boldsymbol{A}}{\partial t} = \frac{\partial\phi}{\partial t}\frac{\mathrm{d}\boldsymbol{A}}{\mathrm{d}\phi} = -\omega\frac{\mathrm{d}\boldsymbol{A}}{\mathrm{d}\phi} \quad \text{and} \quad \frac{\partial\boldsymbol{A}}{\partial z} = \frac{\partial\phi}{\partial z}\frac{\mathrm{d}\boldsymbol{A}}{\mathrm{d}\phi} = k\frac{\mathrm{d}\boldsymbol{A}}{\mathrm{d}\phi}.$$

We thus find

$$\frac{\partial\boldsymbol{A}}{\partial t} = -\frac{\omega}{k}\frac{\partial\boldsymbol{A}}{\partial z} = -c\frac{\partial\boldsymbol{A}}{\partial z}.$$

In addition,

$$\gamma\boldsymbol{v}\cdot\frac{\mathrm{d}\boldsymbol{p}}{\mathrm{d}t} = \gamma\boldsymbol{v}\cdot\frac{\mathrm{d}(m\gamma\boldsymbol{v})}{\mathrm{d}t} = \frac{m}{2}\frac{\mathrm{d}}{\mathrm{d}t}\left[(\gamma v)^2\right] = mc^2\gamma\frac{\mathrm{d}\gamma}{\mathrm{d}t} \quad (11.17)$$

since, from the definition of γ, we have $\gamma^2 v^2 = c^2(\gamma^2 - 1)$. The combination of Eqs (11.16) and (11.17) leads to the determination of the other constant of motion:

$$\frac{\mathrm{d}}{\mathrm{d}t}(\gamma v_z) = c\frac{\mathrm{d}\gamma}{\mathrm{d}t} \implies \Omega = k\gamma(v_z - c) = \text{const.}$$

(4) To answer this question, let us establish the link between the total derivative of the phase ϕ with respect to the proper time τ and the longitudinal component of the momentum $p_z = m\gamma\dot{z}$:

$$\frac{\mathrm{d}\phi}{\mathrm{d}\tau} = k\gamma\dot{z} - \omega\gamma = \Omega, \quad (11.18)$$

whence

$$\phi(\tau) = \Omega\tau + B. \quad (11.19)$$

Initially, $\phi(0,0) = 0$ and $\tau = 0$; we thus obtain $B = 0$. Therefore

$$\Omega = k\gamma(v_0 - c) = -\omega\gamma\left(1 - \frac{v_0}{c}\right),$$

and we find that the phase and the proper time are proportional.

(5) To obtain the explicit form of the trajectories of the electron, we use the two constants of motion, κ_1 and Ω, that we have determined previously. First, we consider a circular polarization ($a_x = a_y = a$). In this case, the equation of motion along z reads

$$\frac{dp_z}{dt} = q\left(v_x\frac{\partial A_x}{\partial z} + v_y\frac{\partial A_y}{\partial z}\right) = kqa(-v_x\sin\phi + v_y\cos\phi). \quad (11.20)$$

As mentioned in the text, we assume that $\kappa_1 = 0$, which gives the expression for the transverse components of the velocity:

$$v_x = -\frac{qa}{m\gamma}\cos\phi \quad \text{and} \quad v_y = -\frac{qa}{m\gamma}\sin\phi. \quad (11.21)$$

Combining Eqs (11.20) and (11.21), we get

$$\frac{dp_z}{dt} = \frac{kq^2a^2}{m\gamma}(\cos\phi\sin\phi - \sin\phi\cos\phi) = 0.$$

This relation, combined with Eq. (11.19), implies that γ is constant. We deduce that $p_z = m\gamma v_z = m\gamma v_0$, or in other words that the longitudinal velocity remains constant $v_z = v_0$. To deduce the explicit expression of the relativistic parameter we write γ as

$$\gamma^2 = 1 + \frac{\gamma^2 v_\perp^2}{c^2} + \frac{\gamma^2 v_z^2}{c^2} = 1 + \frac{\gamma^2 v_0^2}{c^2} + \frac{p_\perp^2}{m^2c^2}.$$

Using the assumption $\kappa_1 = 0$ we have $p_\perp^2 = q^2 A_\perp^2 = q^2 a^2$, and we eventually get

$$\gamma = \left(\frac{1 + q^2a^2/(m^2c^2)}{1 - v_0^2/c^2}\right)^{1/2}. \quad (11.22)$$

The relativistic regime can therefore be reached for $(qa/mc)^2$ sufficiently large, regardless of the value of the initial longitudinal velocity. As expected, we find that the more intense the electromagnetic wave, the more relativistic the dynamics of the electron. It is instructive to rewrite the quantity $(qa/mc)^2$ in terms of the electric field. From $E \sim \omega A$, we get:

$$\frac{q^2a^2}{m^2c^2} \sim \left(\frac{q^2E^2}{m\omega^2}\right)\frac{1}{mc^2}.$$

We find that this quantity is the ratio between the ponderomotive energy (i.e. the wiggling energy of the electron in the oscillatory electric field of the wave) and the rest energy of the electron. When the energy stored in the oscillatory motion of the electron is on the order of is mass energy, the motion becomes relativistic.

The transverse motion is governed by the conservation of the quantity $(m\gamma v_\perp + qA_\perp) = 0$:

$$\frac{dx}{dt} = -\frac{qa}{m\gamma}\cos\phi \Longrightarrow \frac{dx}{d\tau} = -\frac{qa}{m}\cos(\Omega\tau),$$

$$\frac{dy}{dt} = -\frac{qa}{m\gamma}\sin\phi \Longrightarrow \frac{dy}{d\tau} = -\frac{qa}{m}\sin(\Omega\tau),$$

and thus

$$x(\tau) = -\frac{qa}{m\Omega}\sin(\Omega\tau)$$

$$y(\tau) = -\frac{qa}{m\Omega}(1 - \cos(\Omega\tau)).$$

As in the classical regime, the trajectory is a circle. The rotating phase, $\Omega\tau$ is simply Doppler-shifted by the contribution of the longitudinal velocity, v_0. If the initial longitudinal velocity is equal to zero, the equations of motion are identical to those obtained using non-relativistic mechanics.

Note that, like any accelerated charged particle, the electron in the field of the plane wave radiates. Strictly speaking, this damping effect has to be taken into account self-consistently. Here, we have neglected this *radiation reaction* effect to keep the discussion simple.

(6) For a plane wave linearly polarized along y, the y-component of the velocity is, as previously, $v_y = -(qa/m\gamma)\sin\phi$, from which we deduce $y(\tau) = -qa/(m\Omega)(1 - \cos(\Omega\tau))$. The equation of motion for the longitudinal component is

$$\frac{dp_z}{dt} = qv_y\frac{\partial A_y}{\partial z} = qv_yak\cos\phi$$

$$\Longrightarrow \frac{dp_z}{d\tau} = -\frac{kq^2a^2}{2m}\sin(2\phi) = -\frac{kq^2a^2}{2m}\sin(2\Omega\tau)$$

$$\Longrightarrow p_z = m\frac{dz}{d\tau} = \frac{q^2a^2k}{4m\Omega}(\cos(2\Omega\tau) - 1) + m\gamma_0v_0 = \frac{kq^2a^2}{4m\Omega}(\cos(2\Omega\tau) - \eta).$$

with

$$\eta = 1 - \frac{4m^2\Omega v_0}{q^2a^2k}.$$

We finally find

$$z(\tau) = \frac{kq^2a^2}{4m^2\Omega}\left(\frac{\sin(2\Omega\tau)}{2\Omega} - \eta\tau\right).$$

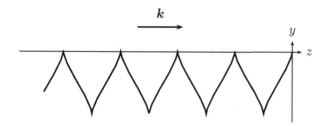

Fig. 11.3 *Trajectory of a relativistic electron subjected to an intense electromagnetic plane wave, linearly polarized along y and propagating along the z axis. The magnetic field induces a longitudinal component of the Lorentz force, that sets the electron into motion along the propagation direction of the wave (actually, the electron moves backwards in the present case).*

The motion is therefore very different from the one obtained with a circularly polarized plane wave, since the particle acquires a longitudinal motion even if, initially, its velocity is zero. This is due to the magnetic field component of the Lorentz force. An example of such a trajectory with no initial velocity ($v_0 = 0$) is depicted in Fig. 11.3.

The relativistic effects described here start to manifest themselves for laser intensities above 10^{18} W $\cdot$ cm^{-2}, which can be achieved experimentally nowadays. Increasing the intensity by several orders of magnitude should allow us in the future to observe, for instance, the creation of electron–positron pairs.

▶ **Further reading.** For an introduction to the physics of relativistic electron–light interaction, the reader may consult the following article: G.A. Mourou, T. Tajima, and S. Bulanov, *Optics in the relativistic regime*, Rev. Mod. Phys. **78**, 309 (2006).

Index